T0214058

Lecture Notes in Computer Science 12633

More information about this subseries at http://www.springer.com/series/7410

Vijay Gadepally · Timothy Mattson ·
Michael Stonebraker · Tim Kraska ·
Fusheng Wang · Gang Luo ·
Jun Kong · Alevtina Dubovitskaya (Eds.)

Heterogeneous Data Management, Polystores, and Analytics for Healthcare

VLDB Workshops, Poly 2020 and DMAH 2020
Virtual Event, August 31 and September 4, 2020
Revised Selected Papers

 Springer

Editors
Vijay Gadepally
Massachusetts Institute of Technology
Lexington, MA, USA

Timothy Mattson
Intel Corporation
Portland, OR, USA

Michael Stonebraker
Massachusetts Institute of Technology
Cambridge, MA, USA

Tim Kraska
Massachusetts Institute of Technology
Cambridge, MA, USA

Fusheng Wang ⓘ
Stony Brook University
Stony Brook, NY, USA

Gang Luo
University of Washington
Seattle, WA, USA

Jun Kong
Georgia State University
Atlanta, GA, USA

Alevtina Dubovitskaya
Lucerne Unviersity of Applied Sciences
Rotkreuz, Switzerland

ISSN 0302-9743   ISSN 1611-3349   (electronic)
Lecture Notes in Computer Science
ISBN 978-3-030-71054-5   ISBN 978-3-030-71055-2   (eBook)
https://doi.org/10.1007/978-3-030-71055-2

LNCS Sublibrary: SL4 – Security and Cryptology

This Springer imprint is published by the registered company Springer Nature Switzerland AG
The registered company address is: Gewerbestrasse 11, 6330 Cham, Switzerland

Preface

In this volume we present the accepted contributions for the VLDB conference workshops entitled: Polystore systems for heterogeneous data in multiple databases with privacy and security assurances (Poly'20) and the Sixth International Workshop on Data Management and Analytics for Medicine and Healthcare (DMAH 2020) held virtually with the 46th International Conference on Very Large Data Bases on August 31 and September 4, 2020.

Poly'20 Overview:

Enterprises are routinely divided into independent business units to support agile operation. However, this leads to "siloed" information systems. Such silos generate a host of problems, such as:

DISCOVERY of relevant data to a problem at hand. For example: Merck has 4000 (+/-) Oracle databases, a data lake, large numbers of files and an interest in public data from the web. Finding relevant data in this sea of information is a challenge.

INTEGRATING the discovered data. Independently constructed schemas are never compatible.

CLEANING the resulting data. A good figure of merit is that 10% of all data is missing or wrong.

ENSURING EFFICIENT ACCESS to resulting data. At scale operations must be performed "in situ", and a good polystore system is a requirement.

It is often said that data scientists spend 80% (or more) of their time on these tasks, and it is crucial to have better solutions.

In addition, the EU has recently enacted GDPR that will force enterprises to assuredly delete personal data on request. This "right to be forgotten" is one of several requirements of GDPR, and it is likely that GDPR-like requirements will spread to other locations, for example California. In addition, privacy and security issues are increasingly an issue for large internet platforms. In enterprises, these issues will be front and center in the distributed information systems in place today.

Lastly, enterprise access to data in practice will require queries constructed from a variety of programming models. A "one size fits all" [1] mentality just won't work in these cases.

DMAH'20 Overview:

The goal of the workshop is to bring together researchers from the cross-cutting domains of research including information management and biomedical informatics. The workshop aims to foster exchange of information and discussions on innovative data management and analytics technologies. We encourage topics that highlight end-to-end applications, systems, and methods addressing problems in healthcare, public health, and everyday wellness; integration with clinical, physiological, imaging, behavioral, environmental, and "omics" data, as well as data from social media and the Web. Our hope for this workshop is to provide a unique opportunity for mutually beneficial and informative interaction between information management and biomedical researchers from interdisciplinary fields.

Organization

POLY'20

Workshop Chairs

Vijay Gadepally	Massachusetts Institute of Technology, USA
Tim Kraska	Massachusetts Institute of Technology, USA
Timothy Mattson	Intel Corporation, USA
Michael Stonebraker	Massachusetts Institute of Technology, USA

Program Committee Members

Danny Weitzner	MIT Internet Policy Research Initiative, USA
Michael Gubanov	Florida State University, USA
Edmon Begoli	Oak Ridge National Laboratory, USA
Dimitris Kolovos	University of York, UK
Amarnath Gupta	University of California, San Diego, USA
Ratnesh Sahay	AstraZeneca, UK
Rada Chirkova	North Carolina State University, USA
Sam Madden	Massachusetts Institute of Technology, USA
Pedro Pedreira	Facebook Inc., USA
Makoto Onizuka	University of Osaka, Japan

DMAH 2020

Workshop Chairs

Fusheng Wang	Stony Brook University, USA
Gang Luo	University of Washington, USA
Jun Kong	Georgia State University, USA
Alevtina Dubovitskaya	Lucerne University of Applied Sciences and Arts and Swisscom, Switzerland

Program Committee Members

Edmon Begoli	Oak Ridge National Laboratory, USA
Yang Cao	Kyoto University, Japan
Blair Christian	Oak Ridge National Laboratory, USA
Dejing Dou	University of Oregon, USA
Alevtina Dubovitskaya	Lucerne University of Applied Sciences and Arts and Swisscom, Switzerland

Peter Elkin	University at Buffalo, USA
Zhe He	Florida State University, USA
Vagelis Hristidis	University of California, Riverside, USA
Athirai Irissappane	University of Washington, USA
Guoqian Jiang	Mayo Clinic College of Medicine and Science, USA
Jun Kong	Georgia State University, USA
Tahsin Kurc	Stony Brook University, USA
Yanhui Liang	Google Inc., USA
Gang Luo	University of Washington, USA
Fusheng Wang	Stony Brook University, USA
Ye Ye	University of Pittsburgh, USA
Rui Zhang	University of Minnesota, USA

Using Demographic Pattern Analysis to Predict COVID-19 Fatalities on the US County Level (Abstract of DMAH 2020 Invited Talk)

Klaus Muller

Stony Brook University, Stony Brook, New York, USA
Akai Kaeru LLC, New York, New York, USA
mueller@cs.stonybrook.edu

Abstract. Unlike pandemics in the past, COVID-19 has hit us in the midst of the information age. We have built vast capabilities to collect and store data of any kind which can be analyzed in myriad ways to help us mitigate the impact of this catastrophic disease. Specifically for COVID-19, data analysis can help local governments to plan the allocation of testing kits, testing stations, and primary care units, and it can help them in setting guidelines for residents, such as the need for social distancing, the use of face masks, and when to open local businesses that enable human contact. Further, it can also lead to a better understanding of pandemics in general and so inform policy makers on the regional and national level. All of this can save both cost and lives. In this tall I will present the results of an ongoing study we conducted using a prominent regularly updated dataset. We used a pattern mining engine we developed to find specific characteristics of US counties that appear to expose them to higher COVID-19 mortality. Furthermore, we also show that these characteristics can be used to predict future COVID-19 mortality.

Bio. Dr. Klaus Mueller received a PhD in computer science from The Ohio State University. He is currently a professor in the Computer Science Department at Stony Brook University and he is also a senior scientist at the Computational Science Initiative at Brookhaven National Lab. He is also a co-founder of Akai Kaeru, the start-up where most of this research took place. His current research interests are explainable AI, visual analytics, and data science. He won the US National Science Foundation Early Career Award, the SUNY Chancellor Award for Excellence in Scholarship and Creative Activity, and the Meritorious Service Certificate and the Golden Core Award of the IEEE Computer Society. Klaus was inducted into the National Academy of Inventors. To date, he has authored more than 200 peer-reviewed journal and conference papers, which have been cited more than 10,000 times. He is a frequent speaker at international conferences, has organized or participated in 18 tutorials on various topics, chaired the IEEE Visualization Conference, and was the elected chair of the IEEE Technical Committee on Visualization and Computer Graphics (VGTC). Klaus currently serves as the Editor-in-Chief of IEEE Transactions on Visualization and Computer Graphics. He is a senior member of the IEEE.

Contents

**Poly 2020: Privacy, Security and/or Policy Issues
for Heterogenous Data**

A Polystore Based Database Operating System (DBOS).   3
*Michael Cafarella, David DeWitt, Vijay Gadepally, Jeremy Kepner,
Christos Kozyrakis, Tim Kraska, Michael Stonebraker,
and Matei Zaharia*

Polypheny-DB: Towards Bridging the Gap Between Polystores
and HTAP Systems.   25
*Marco Vogt, Nils Hansen, Jan Schönholz, David Lengweiler,
Isabel Geissmann, Sebastian Philipp, Alexander Stiemer,
and Heiko Schuldt*

Persona Model Transfer for User Activity Prediction Across
Heterogeneous Domains.   37
Takahiro Hara

PolyMigrate: Dynamic Schema Evolution and Data Migration
in a Distributed Polystore.   42
Alexander Stiemer, Marco Vogt, Heiko Schuldt, and Uta Störl

An Architecture for the Development of Distributed Analytics Based
on Polystore Events.   54
*Athanasios Zolotas, Konstantinos Barmpis, Fady Medhat,
Patrick Neubauer, Dimitris Kolovos, and Richard F. Paige*

Towards Data Discovery by Example   66
*El Kindi Rezig, Allan Vanterpool, Vijay Gadepally, Benjamin Price,
Michael Cafarella, and Michael Stonebraker*

The Transformers for Polystores - The Next Frontier
for Polystore Research.   72
Edmon Begoli, Sudarshan Srinivasan, and Maria Mahbub

DMAH 2020: COVID-19 Data Analytics and Visualization

Open-World COVID-19 Data Visualization [Extended Abstract].   81
Hyunseung Hwang and Steven Euijong Whang

DMAH 2020: Deep Learning based Biomedical Data Analytics

Privacy-Preserving Knowledge Transfer with Bootstrap Aggregation
of Teacher Ensembles   87
  Hong-Jun Yoon, Hilda B. Klasky, Eric B. Durbin, Xiao-Cheng Wu,
  Antoinette Stroup, Jennifer Doherty, Linda Coyle, Lynne Penberthy,
  Christopher Stanley, J. Blair Christian, and Georgia D. Tourassi

An Intelligent and Efficient Rehabilitation Status Evaluation Method:
A Case Study on Stroke Patients.   100
  Yao Tong, Hang Yan, Xin Li, Gang Chen, and Zhenxiang Zhang

Multiple Interpretations Improve Deep Learning Transparency for Prostate
Lesion Detection.   120
  Mehmet A. Gulum, Christopher M. Trombley, and Mehmed Kantardzic

DMAH 2020: NLP Based Learning from Unstructured Data

Tracing State-Level Obesity Prevalence from Sentence Embeddings
of Tweets: A Feasibility Study   141
  Xiaoyi Zhang, Rodoniki Athanasiadou, and Narges Razavian

Enhancing Medical Word Sense Inventories Using Word Sense Induction:
A Preliminary Study   151
  Qifei Dong and Yue Wang

DMAH 2020: Biomedical Data Modelling and Prediction

Teaching Analytics Medical-Data Common Sense   171
  Tomer Sagi, Nitzan Shmueli , Bruce Friedman, and Ruth Bergman

CDRGen: A Clinical Data Registry Generator
(Formal and/or Technical Paper)   188
  Pedro Alves, Manuel J. Fonseca, João D. Pereira,
  and Helena Galhardas

Prediction of lncRNA-Disease Associations from Tripartite Graphs   205
  Mariella Bonomo, Armando La Placa, and Simona E. Rombo

DMAH 2020: Invited Paper

Parameter Sensitivity Analysis for the Progressive Sampling-Based
Bayesian Optimization Method for Automated Machine Learning
Model Selection   213
  Weipeng Zhou and Gang Luo

Short Paper

Extended Abstract: Programming Heterogeneous Data Applications
with Knowledge Graphs.   231
  Michael Cafarella

Author Index   233

Poly 2020: Privacy, Security and/or Policy Issues for Heterogenous Data

A Polystore Based Database Operating System (DBOS)

Michael Cafarella[1], David DeWitt[2], Vijay Gadepally[3(✉)], Jeremy Kepner[3],
Christos Kozyrakis[4], Tim Kraska[2], Michael Stonebraker[2], and Matei Zaharia[5]

[1] University of Michigan, Ann Arbor, USA
[2] MIT, Cambridge, USA
[3] MIT Lincoln Laboratory, Lexington, USA
vijayg@mit.edu
[4] Stanford University, Stanford, USA
[5] Stanford and Databricks, Berkeley, USA

Abstract. Current operating systems are complex systems that were designed before today's computing environments. This makes it difficult for them to meet the scalability, heterogeneity, availability, and security challenges in current cloud and parallel computing environments. To address these problems, we propose a radically new OS design based on *data-centric architecture*: all operating system state should be represented uniformly as database tables, and operations on this state should be made via queries from otherwise stateless tasks. This design makes it easy to scale and evolve the OS without whole-system refactoring, inspect and debug system state, upgrade components without downtime, manage decisions using machine learning, and implement sophisticated security features. We discuss how a database OS (DBOS) can improve the programmability and performance of many of today's most important applications and propose a plan for the development of a DBOS proof of concept.

1 Introduction

Current operating systems have evolved over the last forty years into complex overlapping code bases [4,52,58,71], which were architected for very different environments than exist today. The cloud has become a preferred platform, for both decision support and online serving applications. Serverless computing supports the concept of elastic provision of resources, which is very attractive in many environments. Machine learning (ML) is causing many applications to be redesigned, and future operating systems must intimately support such applications. Hardware is becoming massively parallel and heterogeneous. These "sea changes" make it imperative to rethink the architecture of system software, which is the topic of this paper.

DBOS committee members in alphabetical order. The DBOS Committee, dbos-project@googlegroups.com.

ⓒ Springer Nature Switzerland AG 2021
V. Gadepally et al. (Eds.): Poly 2020/DMAH 2020, LNCS 12633, pp. 3–24, 2021.
https://doi.org/10.1007/978-3-030-71055-2_1

Mainstream operating systems (OSs) date from the 1980s and were designed for the hardware platforms of 40 years ago, consisting of a single processor, limited main memory and a small set of runnable tasks. Today's cloud platforms contain hundreds of thousands of processors, heterogeneous computing resources (including CPUs, GPUs, FPGAs, TPUs, SmartNICs, and so on) and multiple levels of memory and storage. These platforms support millions of active users that access thousands of services. Hence, the OS must deal with a scale problem of 10^5 or 10^6 more resources to manage and schedule. Managing OS state is a much bigger problem than 40 years ago in terms of both throughput and latency, as thousands of services must communicate to respond in near real-time to a user's click [5, 22].

Forty years ago, there was little thought about parallelism. After all, there was only one processor. Now it is not unusual to run Map-Reduce or Apache Spark jobs with thousands of processes using millions of threads [13]. Stragglers creating long-tails inevitably result from substantial parallelism and are the bane of modern systems: incredibly costly and nearly impossible to debug [22].

Forty years ago programmers typically wrote monolithic programs that ran to completion and exited. Now, programs may be coded in multiple languages, make use of libraries of services (like search, communications, databases, ML, and others), and may run continuously with varying load. As a result, debugging has become much more complex and involves a flow of control in multiple environments. Debugging such a network of tasks is a real challenge, not considered forty years ago.

Forty years ago there was little-to-no-thought about privacy and fraud. Now, GDPR [74] dictates system behavior for Personally Identifiable Information (PII) on systems that are under continuous attack. Future systems should build in support for such constructs. Moreover, there are many cases of bad actors doctoring photos or videos, and there is no chain of provenance to automatically record and facilitate exposure of such activity.

Machine learning (ML) is quickly becoming central to all large software systems. However, ML is typically bolted onto the top of most systems as an after thought. Application and system developers struggle to identify the right data for ML analysis and to manage synchronization, ordering, freshness, privacy, provenance, and performance concerns. Future systems should directly support and enable AI applications and AI introspection, including first-order support for declarative semantics for AI operations on system data.

In our opinion, *serverless computing* will become the dominant cloud architecture. One does not need to spin up a virtual machine (VM), which will sit idle when there is no work to do. Instead, one should use an execution environment like Amazon Lambda, Google Cloud Functions, Apache OpenWhisk, IBM Cloud Functions, etc. [18]. As an example, Lambda is an efficient task manager that encourages one to divide up a user task into a pipeline of several-to-many subtasks[1]. Resources are allocated to a task when it is running, and no resources

---

[1] In this paper, we will use Lambda as an exemplar of any resource allocation system that supports "pay only for what you use.".

are consumed at other times. In this way, there are no dedicated VMs; instead there is a collection of short-running subtasks. As such, users only pay for the resources that they consume and their applications can scale to thousands of functions when needed. We expect that Lambda will become the dominant cloud environment unless the cloud vendors radically modify their pricing algorithms. Lambda will cause many more tasks to exist, creating a more expansive task management problem.

Lastly, "bloat" has wrecked havoc on elderly OSs, and the pathlength of common operations such as sending a message and reading bytes from a file are now uncompetitively expensive. One key reason for the bloat is the uncontrolled layering of abstractions. Having a clean, declarative way of capturing and operating on operating system state can help reduce that layering.

These changed circumstances dictate that system software should be reconsidered. In this proposal, we explore a radically different design for operating systems that we believe will scale to support the performance, management and security challenges of modern computing workloads: a *data-centric architecture* for operating systems built around clean separation of all state into database tables, and leveraging the extensive work in DBMS engine technology to provide scalability, high performance, ease of management and security. We sketch why this design could eliminate many of the difficult software engineering challenges in current OSes and how it could aid important applications such as HPC and Internet service workloads. In the next seven sections, we describe the main tenets of this data-centric architecture. Then, in Sect. 9, we sketch a proposal concerning how to move forward.

2 Data-Centric Architecture

One of the main reasons that current operating systems are so hard to scale and secure is the lack of a single, centralized data model for OS state. For example, the Linux kernel contains dozens of different data structures to manage the different parts of the OS state, including a process table, scheduler, page cache, network packet queues, namespaces, filesystems, and many permissions tables. Moreover, each of the kernel components offers different interfaces for management, such as the dozens of APIs to monitor system state (/proc, perf, iostat, netstat, etc.). This design means that any efforts to add capabilities to the system as a whole must be Herculean in scope. For example, there has been more than a decade of effort to make the Linux kernel more scalable on multicores by improving the scalability of one component at a time [10,11,51, 52], which is still not complete. Likewise, it took years to add uniform security management interfaces to Linux – AppArmor [6] and SELinux [65] – that have to be kept in sync with changes to the other kernel components. It similarly took years to enable DTrace [16], a heavily engineered and custom language for querying system state developed in Solaris, to run on other OSs. The OS research community has also proposed numerous extensions to add powerful capabilities to OSs, such as tracing facilities [24,36], tools for undoing changes made by bad actors [46], and new security models [67,79], but these remain academic prototypes due to the engineering cost of integrating them into a full OS.

To improve the scalability, security and operability of OSes, we propose a *data-centric architecture*: designing the OS to explicitly separate data from computation, and centralize all state in the OS into a uniform data model. In particular, we propose using database tables, a simple data model that has been used and optimized for decades, to represent OS state. With the data-centric approach, the process table, scheduler state, flow tables, permissions tables, etc. all become database tables in the OS kernel, allowing the system to offer a uniform interface for querying this state. Moreover, the work to scale or modify OS behavior can now be shared among components. For example, if the OS components access their state via table queries, then instead of reimplementing dozens of data structures to make them scalable on multicores, it is enough to scale the implementations of common table operations. Likewise, new debugging or security features can be implemented against the tabular data model once, instead of requiring separate integration work with each OS component. Finally, making the OS state explicitly isolated also enables radical changes in OS functionality, such as support for zero-downtime updates [3,60], distributed scale-out [7,64], rich monitoring [2,16], and new security models [67,79].

To manage the state in a data-centric operating system, we will require a scalable and reliable implementation of database tables. For this purpose, we simply recommend building the OS over a scale-out DBMS engine, leveraging the decades of engineering and operational experience running mission-critical applications. In other words, we suggest to build a *database operating system (DBOS)*. While the DBMS engine will need some basic resource management functionality to bootstrap its execution, this could be done over a cluster of servers running current OSs, and eventually bootstrapped over the new DBOS. Today, DBMS engines already manage the most critical information in some of the largest computer systems on the planet (e.g. cloud provider control planes). Thus, we believe that they can handle the challenges in a next-generation OS. Moreover, recent trends such as support for polystores [54,69] that combine multiple storage engines will enable the DBMS to use appropriate storage strategies for each of the wide range of data types in an OS, from process tables all the way to file systems.

In more detail, this DBOS approach results in several prescriptive suggestions as discussed in the next section.

2.1 Prescriptive Suggestions

All OS State Should be Stored in Tables in the DBMS. Unix was developed with the mantra that *"everything is a file"*. This mantra should be updated to *"everything is a table"*, with first class support for high performance declarative semantics for query and AI operations on dense, sparse, and hypersparse tables [15,29,33,37,39,44]. For example, there should be a task table with the state of every task known to the system, a flow table with ongoing network flows, a set of tables to represent the file system, etc. [38]

All Changes to OS State Should be Through DBMS Transactions.
The OS will need to include multiple routines in complex imperative code to
implement APIs or complex resource management logic, but when these routines need to access OS state, we will require them to do so through DBMS
transactions. This choice offers several benefits. First, parallelism and concurrency become easier to reason about because there is a transaction manager to
identify conflicts. Second, computation threads in the OS can safely fail without
corrupting system state, enabling a wide range of features including geographic
distribution, improved reliability, and hot-swapping OS code. Third, transactions provide a natural point to enforce security and integrity constraints as is
standard in DBMSs today.

**The DBMS Should be Leveraged to Perform all Functions of Which
It Is Capable.** For example, files should be supported as blobs and tables in
the DBMS. As a result, file operations are simply queries or updates to the
DBMS. File protection should be implemented using DBMS security features
such as view-based access controls for complex security policies. In other words,
there should only be ONE extensible security system, which will hopefully be
better at avoiding configuration errors and leaks than the sprawl of configuration
tools today. Authentication should similarly be done only once using DBMS
facilities. Finally, virtualization and containerization features can elegantly be
implemented using database views: each container simply acts on a view of the
OS state tables restricted to objects in that container.

As a result, ALL system data should reside in the DBMS. To achieve very
high performance, the DBMS must leverage sophisticated caching and parallelization strategies and compile repetitive queries into machine code [2], as is
being done by multiple SQL DBMSs, including Redshift [3]. A DBMS supports
transactions, so ALL OS objects should be transactional. As a result, transactions are implemented just once, and used by everybody.

Decision Support Capabilities Are Facilitated. OSs currently perform
many decision support and monitoring tasks. These include:

- Choosing the next task to run
- Discovering stragglers in a parallel computation
- Finding over(under) loaded resources
- Discovering utilization for the various resources
- Predicting bottlenecks in real-time systems

All of these can be queries to the DBMS.

2.2   Tangible Benefits

Performance Optimization: OS kernel subsystems have often undergone
extensive refactoring to improve performance by changing the data structures
used to manage various state [32, 53, 70, 76]. If the OS had been designed around
a DBMS instead, many of these updates would amount to changing indexes or

changing operator implementations in the DBMS (e.g., adding parallel versions of operators). Moreover, the DBMS approach would enable further methods to improve performance that are not implemented in OSes today, such as cost-based optimization (switching access paths for an operation based on the current data statistics and expected size of the operation) or adaptive mid-query reoptimization.

Security: DBMS access control tools such as view, attribute and role based ACLs [19, 75] can elegantly implement many of the security policies in SELinux, AppArmor and other OS security modules. Moreover, if these rules are implemented as view definitions or SQL statements within the DBMS, the security checking code can be compiled into the queries that regular OS operations run, instead of being isolated in a separate module that adds overhead to OS operations [49].

Virtualization and Containerization: Tremendous engineering effort has gone into enabling virtualization and containerization in OSes over the past decade, i.e., enabling a single instance of the OS to host multiple applications that each get the abstraction of an isolated system environment. These changes have generally required modifying all data structures and a large amount of logic in the kernel to support different "namespaces" of objects for each container. With DBOS, virtualization and containerization can elegantly be achieved using DBMS views: each container's DBMS queries only have access to a view that restricts to objects with that container ID, whereas a root user can have access to all objects. We believe that many queries and logic in OS components would not have had to be modified at all to add virtualization with this approach, other than being made to run on these views instead of on the raw OS state tables.

Geographic Distributability: After all, nodes in a cloud vendor's offering are geographically distributed. Transactional replication is a desired service of cloud offerings. This can be trivially provided by a geographically dispersed DBMS. This is in keeping with "implement any function only once; in the interest of simplicity".

More Sophisticated File Management: Since files are stored in the DBMS, as blobs and tables, and the directory structure is a collection of tables, and SQL access control is used for protection, the large amount of code that implements current file systems, essentially disappears. Also, we claim that current DBMSs which use aggressive compilation query and caching have gotten a great deal faster than the DBMSs of yesteryears. Also, multinode main memory DBMSs such as VoltDB and MemSQL are capable of tens of millions of simple transactions per second. Since a file read/write is just such a simple transaction, we believe that our proposed implementation can be performance competitive. In addition, more sophisticated file search becomes trivial to implement. For example, finding all files underneath a specific directory accessed in the last 24 h that are more than 1GByte in size is merely a SQL query. The net result is additional features, much less code and (hopefully) competitive performance.

Better Scheduling: There will be task and resource tables in the DBMS capturing what tasks runs on cores, chips, nodes, and datacenters and what resources are available. Scheduling thousands of parallel tasks in such environments as Map-Reduce and Spark is mainly an exercise in finding available resources and stragglers, because running time is the time of the slowest parallel task. Finding outliers in a large task table is merely a decision support query that can be coded in SQL. Again, we believe that the additional functionality can be provided at a net savings in code.

Enhanced State Management: Using this approach it is straight-forward to divide application state into two portions. The first is transient and can be stored in data structures external to the DBMS. The second is persistent and must be stored in the DBMS transactionally. Since replication will be provided for all DBMS objects, application failures can merely failover to a new instance. This instance reads the persistent state from the DBMS and resumes the computation. This failover architecture was pioneered by Tandem Computers in the 1980's and can be provided nearly for free using our architecture.

Additional benefits accrue to this architecture by using a modern "serverless" application architecture, a topic which we defer to Sect. 8.

3   Task Communication

Data communications can be readily expressed as operations on a geographically distributed DBMS. A pull-based system can be supported by the sender writing a record into the DBMS and the receiver reading it. A push-based system can be supported by the sender writing to the DBMS and setting a trigger to alert the receiver when he becomes active. This can be readily extended to multiple senders and recipients. In addition, DBMS transactions support exactly-once messages. Such an approach significantly simplifies programming allowing the programmer to easily implement non-blocking send programs that have been demonstrated comparable bandwidth to more complex messaging systems [12,41].

The CPU overhead of conventional TCP/IP communication is considered onerous by most, and new lighter-weight mechanisms, such as RDMA and kernel-bypass systems, are an order of magnitude faster [9,57]. Hence, it seems reasonable to build special purpose lightweight communication systems whose only customer is the DBMS. This has already been shown to accelerate DBMS transactions by an order of magnitude, relative to TCP/IP in a local area networking environment [78], and it is possible that appropriate hardware could offer advantages of this approach in a wide area networking world. As such, it is an interesting exercise to see if a competitive messaging system can be done through the DBMS. It should also be noted that Amazon Lambda uses a storage-based communication system [73]. Of course, a performant implementation would use something much faster than S3, such as a multi-node main memory DBMS.

If this approach is successful, this will lower the complexity of future system software by replacing a heavyweight general purpose system with a lightweight and optimized, special purpose one. It seems highly likely that the approach will work well in a hardware-assisted LAN environment. WAN utilization seems more speculative.

4   GDPR and Privacy Standards

It is clear that privacy will be a future requirement of all system software. GDPR [74] is the European law that mandates "the right to be forgotten". In other words, Personally Identifiable Information (PII) that a service holds on an individual must be permanently removed upon a user request. In addition, data access must be based on the notion of "purposes". Purposes are intended to capture the idea that performing aggregation for reporting purposes is a very different use case than performing targeted advertising based on PII data. In SQL DBMSs access control is based on the notion of individuals and their roles. These constructs have nothing to do with purposes, and a separate mechanism is required. Obviously, this is a DBMS service.

As noted in [30], a clean DBMS design can facilitate locating and deleting PII data inside the DBMS. However, one must also deal with the case where data is copied to an application and then sent to a second application. Since all communication between applications goes through the DBMS, this message can be recorded by the DBMS, allowing the DBMS to track PII data even when it goes out to applications. Of course, this will not prevent a malicious human from writing PII data to the screen and copying it outside of the system. To deal with these kinds of leaks, applications must be "sandboxed" either virtually or cryptographically which can be readily incorporated into the database [26,28, 40,59,61,77].

5   Strong Provenance Guarantees

Data provenance is key to addressing many of the ills of modern data-centric life. Consider the following problems:

Data Forging: Detecting whether a photograph is doctored has become impossible for the typical news consumer. Even if a news service wants to provide trustworthy authorship information about its articles and photos, it has no trustworthy way to do so. Simply signing a photograph at the time it was taken is not sufficient, since there are some data-mutating operations (such as cropping or color adjustment) that news organizations must perform before publication.

Data Debugging: Modern machine learning projects involve huge data pipelines, incorporating datasets and models from many different sources. Debugging pipeline output requires closely examining and testing these different inputs. Unfortunately, these inputs can come from partners with opaque engineering pipelines, or are incorporated in an entirely untracked manner, such as via a downloaded email attachment. As a result, simply enumerating the inputs to a data pipeline can be challenging, and fixing "root cause data problems" is frequently impossible.

Data Spills: Today, an inadvertent data revelation is an irreversible mistake. There is no such thing as cleaning up after a database of social security numbers is mistakenly posted online. Although data handling practices must and can be improved, ensuring total data privacy today is a very difficult and brittle problem.

Data Consumption and Understanding: Much of modern life (as a professional, a consumer, and a citizen) consists of consuming and acting on data. The data processes that produce human-comprehensible outputs, such as the plots in a scientific article, are so complicated that it is quite easy for there to be errors that are undetectable even to the producer. Consider the case of economists Carmen Reinhart and Kenneth Rogoff, who in 2010 wrote an enormously influential article on public finance, cited by Representative Paul Ryan to defend a 2013 budget proposal, that was later found to be based on simplistic errors in an Excel spreadsheet [72]. The authors did not acknowledge the error until three years after the paper was first written. Responsible data use means people must be able to quickly examine and understand the processes that yield the data artifacts all around us.

Data Policy Compliance: Datasets and models often carry policies about how they can be used. For example, a predictive medical model might be appropriate for some age populations, but not others. Unfortunately, it is impossible for anyone, whether a data artifact producer or consumer, to have confidence about how data is being used.

A strong data provenance system would help address all of the above problems. All data operations by a modern operating system, such as copying, mutating, transmitting, etc., should be tracked and stored for possible later examination. It should be impossible to perform operations on a modern OS that sidestep responsible data provenance tracking. Our proposed DBOS architecture effectively logs all such operations, allowing an authoritative chain of provenance to be recorded. (As with all the data the system collects, it will be stored in a DBMS.) This will support solutions to all of the above issues, requiring only log processing applications. Furthermore, first-class support for provenance throughout OS data structures will also simplify many system administration tasks, such as recovering from user errors or security breaches [20].

6   Self-adaptive via Modern ML

Designing an operating system requires making assumptions about its future workload and data. These assumptions then materialize themselves as default parameters, heuristics, and various compromises. Unfortunately, all these decisions can significantly impact performance, especially if the assumptions turn out to be wrong. For example, if we assume that the OS mainly runs very short

Lambda-like functions, then reducing the overhead of starting a Lambda function may be more critical than optimal scheduling. However, if we assume the workload is dominated by long-running memory intensive services, we require a very different scheduling algorithm, fair resource allocation strategies, and service migration techniques, whereas the startup time will matter very little.

Moreover, operating systems offer a variety of knobs to tune the system for a particular workload or hardware. While providing flexibility, all the options put a burden on the administrator to set the knobs correctly and to adjust them in the case the workload, data, or hardware changes.

To overcome those challenges, we suggest that DBOS should be introspective, adaptable, and self-tuning through two design principles:

Knob-Free Design: We believe that all parameters of the system should be designed to be self-tuning from the beginning. That is, DBOS will deploy techniques similar to SmartChoices [17] for all parameters and constants to make them automatically tuneable. The key challenge in globally optimizing all these parameters is then to gather and analyzing the state of the OS and the different components. Storing all this information in the OS database will significantly simplify the process and make true self-tuning possible.

Learned Components: To address a wide range of use cases, the system developer often has to make algorithmic compromises. For instance, every operating system requires a scheduling algorithm, but the chosen scheduling algorithm might not be optimal under all workloads or hardware types. In order to provide the best performance, we envision that the system is able to automatically switch the algorithm used, based on the workload and data. This would apply to scheduling, memory management, etc. [21,23].

In some cases it might be even possible to learn the entire component or parts of it. For example, recent results have shown that it is sometimes possible to learn a scheduling algorithm, which performs better than traditional more static heuristics [55,56]. This learning of components would allow the system to more readily adapt to the workload and data, and perhaps provide unprecedented performance.

To achieve a knob-free design and learned components, we suggest that the DBOS needs to be designed from the beginning to be Reinforcement Learning (RL)-enabled. RL is the leading technique to tune knobs and build components based on the observed behaviour in an online fashion. Today, RL is usually added as an afterthought. This leads to several problems including difficulty in finding the right award function or supporting the required RL exploration phase. In many cases this requires the extra work of building a simulator or a light-weight execution environment to try out new approaches. By making RL a first-class citizen in the system design, we believe that we can overcome these challenges. Moreover, managing all state data in a database and making it analyzable, will again be a key enabler for this effort.

If successful, the resulting system would be able to quickly adapt itself to changing conditions and provide unprecedented performance for a wide range of workloads while making the administration of the system considerably easier.

7   Diverse Heterogenous Hardware

Managing compute, storage, and communication hardware is a primary function for an operating system. The key abstractions in existing operating systems were developed for the homogeneous hardware landscape of the last century. Kernel threads (processes), virtual memory, files, and sockets were sufficient to abstract and manage single-core computers with limited main memory backed by a slow hard disk, connected with low-bandwidth, high latency networking.

Present-day hardware looks radically different. A single server machine contains tens to hundreds of cores in one or more chips, terabytes of main memory across a dozen channels, and multiple storage devices (SSDs and HDDs). The end of Dennard scaling [50] and the ascent of machine learning applications has led to the introduction of domain-specific accelerators like GPUs and TPUs, each with its own primitives for massively parallel computation and high-bandwidth memory [34]. The end of scaling for DRAM technology is motivating multi-level main memory systems using storage-class memories (SCM) [35]. Network interfaces allow direct access to remote memory at speeds faster than local storage. Beyond the single node, concepts such as multi-cloud, edge cloud, globally replicated clouds, and hardware disaggregation introduce heterogeneity in the type and scale of hardware resources. Existing operating systems were not designed for such scales or heterogeneity. This shortcoming is a primary culprit for the software bloat in applications and operating systems, including kernel bypass subsystems. Solutions have limited portability and are difficult to understand, debug, and reuse.

Placing the operating system state in a DBMS introduces two properties that are useful in managing heterogeneous hardware. First, it clearly separates compute from data access. The operating system can manage data placement, caching, replication, and synchronization separately from the accelerated functions that operate on it. Second, it clearly separates control-plane from data-plane actions. One can improve or customize control-plane operations, such as scheduling, independently of the compute implementation using the best available accelerators.

To run efficiently on heterogeneous hardware, DBOS will be designed around two key principles.

Accelerated Interfaces to DBMS: DBOS will implement the interfaces that allow heterogeneous hardware to interact with the DBMS, hiding the overall system scale and complexity. For example, the interface to a compute accelerator like a TPU can be a query that applies a user-defined function (UDF). The accelerator implements the UDF, while DBOS implements the query that involves preparing inputs and outputs. This interface remains constant regardless if the accelerator is local, disaggregated, or in a remote datacenter. The accelerator state is stored in the DBMS to facilitate scheduling and introspection. DBOS will directly manage memory and storage layers, as part of the DBMS resources available for data sharing, replication, or caching. DBOS interfaces will leverage existing hardware mechanisms, such as virtual memory, as well as emerging mechanisms such as zero-copy/direct memory access networking interfaces or coherent fabrics (CXL). Over the time, hardware mechanisms will evolve to further accelerate the interactions between the DBMS and heterogeneous hardware. For example, SmartNICs will be optimized to accelerate DBMS interfaces, not just RDMA protocols, while GPUs and TPUs will directly support DBMS data operations.

Accelerating the DBMS Itself: The performance and scalability of DBOS itself relies heavily on the speed of DBMS operations. In addition to distributed execution and extensive caching, the DBMS will build upon modern hardware – accelerators, storage class memory, and fast SmartNICs. Since all communication, dataplane, and control plane operations interface with the DBMS, the deployment of specialized accelerators for common DB operations like joins, filters, and aggregations will likely become essential [1].

8   Programming Model

Historically, the programming model of choice was a single-threaded computation with execution interspersed with stalls for I/O or screen communication. This model effectively requires multi-tasking to fill in for the stalls. In turn, this requires interprocess protection and other complexity.

Instead, we would recommend that everybody adopt the Lambda model, popularized by AWS [73]. In other words, computation is done in highly parallel "bursts", and resources are relinquished between periods of computation [25]. This model allows one to give the CPU to one task at a time, eschewing multithreading and multiprogramming. In addition, parallel processing can be done with a collection of short-lived, stateless tasks that communicate through the DBMS. The DBMS optimizes the communication by locally caching and coscheduling communicating tasks when possible. In effect, this is "server-less computing", whereby one only pays for resources that are used and not for long-lived tasks. Hence, under current cloud billing practices, this will save significant dollars.

That means DBOS should adopt the Lambda model as well. One should divide up a query plan into "steps" (operators). Each operator is executed (in parallel) and then dies. State is recorded in the DBMS. Sharding of the data allows operator parallelism.

Each Lambda task is given a exclusive set of resources, i.e., one or more cores until it dies. In the interest of simplicity and security, multi-tenancy and multi-threading may be turned off.

There is a sharded scheduling table in the DBMS. A task is runnable or waiting. The scheduler picks a runnable task—via a query—and executes it. When the task quits, the scheduler loops. This will work well as long as applications utilize the Lambda model.

Dynamic optimization in the OS is gated by the time it takes stop, checkpoint, migrate, and restart applications/processes/threads. In the cloud, this is often minutes, which means that very little dynamic optimization is possible. Recent work has demonstrated that hand-coded fast launch (thousands of applications per second) is possible [62,63]. This is all human-controlled static optimization [14]. The optimizing scheduler in DBOS should be able to do this dynamically and launch millions of applications per second [38].

9   Plans for a Proof of Concept

Obviously, DBOS is a huge undertaking. An actual commercial implementation will take tens of person-years. As such, we need to quickly validate the ideas in this document. Hence, we discuss demonstrating the validity of the ideas and then discuss convincing the systems community that DBOS is worth the effort involved (Fig. 1).

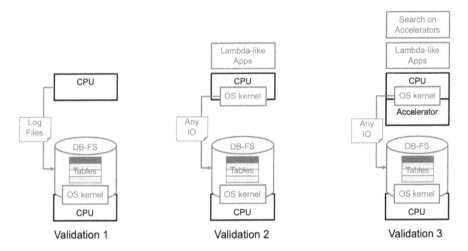

Fig. 1. The three stages of our intended proof of concept. Phase 1 comprises a proof of concept to demonstrate database performance. Phase 2 includes work necessary for log processing. Phase 3 shows how we can manage accelerators and implement end-to-end microservice applications.

9.1   Key Characteristics

A key challenge is to show a DBMS capable of acceptable performance and scalability to form the foundation of DBOS. We believe that such a system should have the following characteristics:

Multi-core, Multi-node Executor: Many DBMSs support this today.

Server-Less Architecture: Commercial DBMSs are moving toward allocating CPU resources on a per-query basis. Snowflake has moved aggressively in this direction, based on a distributed file system (S3) aggressive caching and sharding only for CPU resources [66].

Polystore Architecture: Clearly, DBOS will need to manage data from heterogenous sources such as process tables, schedulers, network tables, namespaces, and many permissions tables. It is likely that any single data management system will be able to efficiently manage the diversity and scale of the associated data structures. Different OS functionality will naturally fit into different types of storage engines and a polystore architecture [54,69] can provide a single interface to these disparate and federated systems. A critical system characteristic would be to avoid developing a "one size fits all" [27,45,68] solution that is incapable of adapting as new types of data are collected and managed by DBOS.

Open Source Code: Obviously any code in a DBOS prototype should be readily available.

Lambda-Style, Serverless Runtime System: This will facilitate optimizing resource allocation.

Possible choices include SciDB, Presto, Accumulo, etc. We think the best option is to start with a prototype that comprises a DBMS built on an MIT Lambda-style system.

We view the key design choices of AWS Lambda as reservation-free, fixed-resource service for short-lived functions and will embody those in our own system. Other choices in today's commercial version of Lambda, such as S3 as the exclusive storage system, or the lack of direct communication between functions, seem like they should be rethought. It is unclear whether uniform resource constraints on the Lambda functions is a key design choice, or whether the system should offer heterogeneous resource constraints to enable a more flexible development environment.

We would expect to replace S3 as the storage system with something much faster [47,48], based on the discussion earlier. We expect in one person year, we could demonstrate a LAN-based system along these lines. We would then expect to test the performance of this prototype in two contexts. The first goal is to provide file system performance comparable to today's systems. In addition, we expect to show our communication implementation can be comparable or faster to traditional TCP/IP networking.

To bootstrap running the DBMS itself, we plan to rely on minimal operating systems that have already been designed for cloud environments, such as unikernels, Dune or IX [8,9], which are designed to run one application at a time and to give it high-performance access to the hardware. We will also make sure that the DBMS runs on Linux systems for easy development. The main facilities that the DBMS needs to bootstrap are a boot and configuration process, network access (which can also be used for logging), threads, and an interface to access storage. In the latter case, because the DBMS will manage all large data structures, raw block access may be sufficient. Today's minimal OSes already support these facilities for hosting server applications as efficiently as possible in virtualized datacenters.

9.2   Demonstration of Utility: Log Processing

As a first example of using DBOS to improve current OS functionality, we will implement a data-centric log processing and monitoring infrastructure in DBOS that can monitor applications using existing OSes such as Linux. OSes, Networks, Schedulers, and File Systems generate enormous amounts of logs and metadata which are mostly kept in raw files. Attempts to put these in databases (OS logs to Splunk; Network logs to NetAPP; Scheduler logs to MySQL; File System metadata to MySQL) barely meet minimal auditing requirements.

A DBMS-based OS that organically stored these data in a high-performance database with first class support for dense, sparse, and hypersparse tables would be a huge win as it would make these data readily analyzable and actionable. It would also be able to execute streaming queries to compute complicated monitoring views in real time in order so simplify system management; simple metrics such as "how many files has each user created" can sometimes take hours to run with today's file systems and OSes. Our team has conducted experiments showing the high-performance databases such as Apache Accumulo, SciDB, and RedisGraph can easily absorb this data while enabling analysis that are not currently possible [15, 38, 42, 43]. For example, "All files touched by a user during a time window", "Largest 10 folders owned by a user", "Computing cycles consumed by an application during a time window", "Network traffic caused by a specific application", ... These are very important questions for Cloud operators and very difficult to answer and require custom built tools to do so. A DBMS OS should be able to answer these questions by design.

9.3   Demonstration of Utility: Managing Accelerators

In Sect. 7, we discussed DBMS support for heterogeneous hardware, GPUs and FPGAs, based on user-defined DBMS functions. Our plan is to implement a prototype of this functionality to demonstrate its feasibility and performance.

One of the defining features of modern datacenters is hardware heterogeneity. Far from being a uniform pool of machines, datacenters offer machines with different memory, storage, processing, and other capacities. Most notably, different machines offer vastly different accelerator capacities. Although GPUs for machine learning tasks comprise the most common class of accelerator, datacenters also contain FPGAs and other accelerators for video processing and encryption applications. These accelerators can be expensive: it is not feasible to outfit every machine in a large system with a top-flight GPU. Matching a heterogeneous workload to a heterogeneous pool of resources is a complicated and important task that is tailor-made for machine- rather than human-driven optimization.

To address this challenge, we need to first design the DBMS-based API in DBOS to allow for portability. The same user code can drive execution on a local or remote GPU. Next, we need to exploit the flexibility of Lambda-style task allocation and the visibility into system state through the DBMS in order to implement scheduling algorithms that better utilize the datacenter resources of naive server-centric allocation schemes. We will demonstrate this functionality by running a range of workloads on small clusters and by simulating larger, datacenter environments.

9.4   Demonstration of Utility: End-to-end Microservice Applications

Since DBOS is designed around a distributed DBMS, it is a natural fit for data mining applications like the log processing discussed in Sect. 9.2. However, it is not as obvious a match for online-serving applications, such as social networks, e-commerce sites, and media services, that consume large fractions on cloud systems. These applications consist of tens to thousands of microservices that must quickly communicate and respond to user actions within tight service level objectives (SLOs) [31]. Some microservices are simple tasks, such as looking up session information, while others are complicated functions such as recommendation systems based on neural networks or search functions using distributed indices. Microservice applications form the bulk of software-as-a-service products today and are the most critical operational applications for many organizations.

We will prototype an end-to-end microservices workload, such as a Twitter-like social network, in order to evaluate DBOS's feasibility for these applications. During this process, we will answer two key questions. First, can DBOS support the computation and communication patterns of such latency-critical applications in a performant manner? Second, can DBOS help address the challenges in developing, scaling, and evolving such applications over time?

With DBOS, a social network will be implemented as a collection of serverless functions operating on multiple database tables. This presents multiple opportunities for performance optimization. For example, DBOS can colocate communication functions to avoid remote communication, or selectively introduce new caching layers and indexes. Accelerators are also now used in many components of microservice applications, such as recommendation engines for social network content and search result re-ranking, so we will use the accelerator management capabilities in Sect. 9.3 to automatically offload and optimize these tasks.

Finally, because DBOS uses a serverless model, data management decisions such as sharding and replicating datasets or evolving schemas are separated from the application code. This makes it significantly easier for application developers to implement architectural changes that are very difficult in microservice applications today. We will show how to use DBOS to easily implement several such architectural changes:

1. Changing the partitioning and schema of data in the application to improve performance (a common type of change that requires large engineering efforts in today's services).
2. Changing the partitioning of compute logic, e.g., moving from a "monolith" of co-located functions to separately scaling instances for different parts of the application logic.
3. Making the application GDPR-compliant, by storing each user's data in their geographic region and using the data provenance features of DBOS to track which data was derived from each user or delete it on-demand.

4. Changing the security model (*e.g.*, which users can see data from minors or from European citizens) without having to refactor the majority of application code.

10   Conclusions

We have presented a dramatically simpler view of systems software which avoids implementing the same functions in multiple components. Instead, the architecture bets on a sophisticated DBMS to implement most functionality. Section 9 suggested initial experiments to demonstrate feasibility. Obviously, these steps should be carried out first.

Following that, there are still many unanswered questions. The most notable one is "Can this scale to a million nodes?". To the best of our knowledge, nobody has built a distributed DBMS at this scale. Clearly, there will be unforeseen bottlenecks and inefficiencies to contend with. Managing storage for a 1M node DBMS will be a challenge. A second question is "Can this be built to function efficiently?" Section 9 discussed the file system and IPC. However, memory management, caching, scheduling, and outlier processing are still issues. Obviously, the next step is to build a full-function prototype to answer these questions.

Acknowledgments. This work was partially supported by National Science Foundation CCF-1533644 and United States Air Force Research Laboratory Cooperative Agreement Number FA8750-19-2-1000. Any opinions, findings, conclusions or recommendations expressed in this material are those of the author(s) and do not necessarily reflect the views of the National Science Foundation or the United States Air Force. The U.S. Government is authorized to reproduce and distribute reprints for Government purposes notwithstanding any copyright notation herein. The authors would also like to thank Charles Leiserson, Peter Michaleas, Albert Reuther, Michael Jones, and the MIT Supercloud Team.

References

1. Agrawal, S.R., et al.: A many-core architecture for in-memory data processing. In: Proceedings of the 50th Annual IEEE/ACM International Symposium on Microarchitecture, MICRO-50 2017, pp. 245–258. Association for Computing Machinery, New York (2017). https://doi.org/10.1145/3123939.3123985
2. Ardelean, D., Diwan, A., Erdman, C.: Performance analysis of cloud applications. In: 15th USENIX Symposium on Networked Systems Design and Implementation (NSDI 2018), pp. 405–417. USENIX Association, Renton (2018). https://www.usenix.org/conference/nsdi18/presentation/ardelean
3. Arnold, J., Kaashoek, M.F.: Ksplice: Automatic rebootless kernel updates. In: Proceedings of the 4th ACM European Conference on Computer Systems, EuroSys 2009, pp. 187–198. Association for Computing Machinery, New York (2009). https://doi.org/10.1145/1519065.1519085
4. Atlidakis, V., Andrus, J., Geambasu, R., Mitropoulos, D., Nieh, J.: POSIX abstractions in modern operating systems: the old, the new, and the missing. In: Proceedings of the Eleventh European Conference on Computer Systems, EuroSys 2016.

Association for Computing Machinery, New York (2016). https://doi.org/10.1145/2901318.2901350

5. Barroso, L., Marty, M., Patterson, D., Ranganathan, P.: Attack of the killer microseconds. Commun. ACM **60**(4), 48–54 (2017). https://doi.org/10.1145/3015146

6. Bauer, M.: Paranoid penguin: an introduction to Novell AppArmor. Linux J. **2006**(148), 13 (2006)

7. Baumann, A., et al.: The multikernel: a new OS architecture for scalable multicore systems. In: Proceedings of the ACM SIGOPS 22nd Symposium on Operating Systems Principles, SOSP 2009, pp. 29–44. Association for Computing Machinery, New York (2009). https://doi.org/10.1145/1629575.1629579

8. Belay, A., Bittau, A., Mashtizadeh, A., Terei, D., Mazières, D., Kozyrakis, C.: Dune: Safe user-level access to privileged CPU features. In: Proceedings of the 10th USENIX Conference on Operating Systems Design and Implementation, OSDI 2012, pp. 335–348. USENIX Association, USA (2012)

9. Belay, A., Prekas, G., Klimovic, A., Grossman, S., Kozyrakis, C., Bugnion, E.: IX: A protected dataplane operating system for high throughput and low latency. In: 11th USENIX Symposium on Operating Systems Design and Implementation (OSDI 2014), pp. 49–65. USENIX Association, Broomfield (2014). https://www.usenix.org/conference/osdi14/technical-sessions/presentation/belay

10. Bhat, S.S., Eqbal, R., Clements, A.T., Kaashoek, M.F., Zeldovich, N.: Scaling a file system to many cores using an operation log. In: Proceedings of the 26th Symposium on Operating Systems Principles, SOSP 2017, pp. 69–86. Association for Computing Machinery, New York (2017). https://doi.org/10.1145/3132747.3132779

11. Boyd-Wickizer, S., et al.: An analysis of linux scalability to many cores. In: Proceedings of the 9th USENIX Conference on Operating Systems Design and Implementation, OSDI 2010, pp. 1–16. USENIX Association, USA (2010)

12. Byun, C., et al.: Large scale parallelization using file-based communications. In: 2019 IEEE High Performance Extreme Computing Conference (HPEC), pp. 1–7 (2019)

13. Byun, C., et al.: LLMapReduce: multi-level map-reduce for high performance data analysis. In: 2016 IEEE High Performance Extreme Computing Conference (HPEC), pp. 1–8 (2016)

14. Byun, C., et al.: Optimizing Xeon phi for interactive data analysis. In: 2019 IEEE High Performance Extreme Computing Conference (HPEC) (2019). https://doi.org/10.1109/hpec.2019.8916300

15. Cailliau, P., Davis, T., Gadepally, V., Kepner, J., Lipman, R., Lovitz, J., Ouaknine, K.: RedisGraph graphBLAS enabled graph database. IEEE (2019). https://doi.org/10.1109/ipdpsw.2019.00054

16. Cantrill, B.M., Shapiro, M.W., Leventhal, A.H.: Dynamic instrumentation of production systems. In: Proceedings of the Annual Conference on USENIX Annual Technical Conference, ATEC 2004, p. 2. USENIX Association, USA (2004)

17. Carbune, V., Coppey, T., Daryin, A., Deselaers, T., Sarda, N., Yagnik, J.: SmartChoices: hybridizing programming and machine learning. In: Reinforcement Learning for Real Life (RL4RealLife) Workshop in the 36th International Conference on Machine Learning (ICML) (2019). https://arxiv.org/abs/1810.00619

18. Castro, P., Ishakian, V., Muthusamy, V., Slominski, A.: The rise of serverless computing. Commun. ACM **62**(12), 44–54 (2019)

19. Chamberlin, D.D., et al.: A history and evaluation of system R. Commun. ACM **24**(10), 632–646 (1981). https://doi.org/10.1145/358769.358784

20. Chandra, R., Kim, T., Zeldovich, N.: Asynchronous intrusion recovery for interconnected web services, pp. 213–227 (2013). https://doi.org/10.1145/2517349.2522725
21. Cortez, E., Bonde, A., Muzio, A., Russinovich, M., Fontoura, M., Bianchini, R.: Resource central: understanding and predicting workloads for improved resource management in large cloud platforms. In: Proceedings of the 26th Symposium on Operating Systems Principles, SOSP 2017, pp. 153–167. Association for Computing Machinery, New York (2017). https://doi.org/10.1145/3132747.3132772
22. Dean, J., Barroso, L.A.: The tail at scale. Commun. ACM **56**(2), 74–80 (2013). https://doi.org/10.1145/2408776.2408794
23. Delimitrou, C., Kozyrakis, C.: Paragon: QoS-aware scheduling for heterogeneous datacenters. In: Proceedings of the Eighteenth International Conference on Architectural Support for Programming Languages and Operating Systems, ASPLOS 2013, pp. 77–88. Association for Computing Machinery, New York (2013). https://doi.org/10.1145/2451116.2451125
24. Feiner, P., Brown, A.D., Goel, A.: Comprehensive kernel instrumentation via dynamic binary translation. In: Proceedings of the Seventeenth International Conference on Architectural Support for Programming Languages and Operating Systems, ASPLOS XVII, pp. 135–146. Association for Computing Machinery, New York (2012). https://doi.org/10.1145/2150976.2150992
25. Fouladi, S., et al.: From laptop to lambda: outsourcing everyday jobs to thousands of transient functional containers. In: 2019 USENIX Annual Technical Conference (USENIX ATC 2019), pp. 475–488. USENIX Association, Renton (2019). https://www.usenix.org/conference/atc19/presentation/fouladi
26. Fuller, B., et al.: SoK: cryptographically protected database search, pp. 172–191 (2017)
27. Gadepally, V., et al.: The bigDAWG polystore system and architecture, pp. 1–6 (2016)
28. Gadepally, V., et al.: Computing on masked data to improve the security of big data, pp. 1–6 (2015)
29. Gadepally, V., et al.: D4M: Bringing associative arrays to database engines, pp. 1–6 (2015)
30. Gadepally, V., et al.: Heterogeneous Data Management, Polystores, and Analytics for Healthcare: VLDB 2019 Workshops, Poly and DMAH, Revised Selected Papers, vol. 11721. Springer Nature (2019)
31. Gan, Y., et al.: An open-source benchmark suite for microservices and their hardware-software implications for cloud & edge systems. In: Proceedings of the Twenty-Fourth International Conference on Architectural Support for Programming Languages and Operating Systems, ASPLOS 2019, pp. 3–18. Association for Computing Machinery, New York (2019). https://doi.org/10.1145/3297858.3304013
32. Gleixner, T.: Refactoring the Linux kernel (2017). https://kernel-recipes.org/en/2017/talks/refactoring-the-linux-kernel/
33. Hutchison, D., Kepner, J., Gadepally, V., Fuchs, A.: Graphulo implementation of server-side sparse matrix multiply in the accumulo database. In: 2015 IEEE High Performance Extreme Computing Conference (HPEC), pp. 1–7 (2015)
34. Jouppi, N.P., et al.: A domain-specific supercomputer for training deep neural networks. Commun. ACM **63**(7), 67–78 (2020). https://doi.org/10.1145/3360307
35. Kamath, A.K., Monis, L., Karthik, A.T., Talawar, B.: Storage class memory: principles, problems, and possibilities (2019)

36. Kedia, P., Bansal, S.: Fast dynamic binary translation for the kernel. In: Proceedings of the Twenty-Fourth ACM Symposium on Operating Systems Principles, SOSP 2013, pp. 101–115. Association for Computing Machinery, New York (2013). https://doi.org/10.1145/2517349.2522718
37. Kepner, J., et al.: Mathematical foundations of the graphBLAS, pp. 1–9 (2016)
38. Kepner, J., et al.: TabulaROSA: tabular operating system architecture for massively parallel heterogeneous compute engines. In: 2018 IEEE High Performance extreme Computing Conference (HPEC), pp. 1–8 (2018)
39. Kepner, J., et al.: Associative array model of SQL, NoSQL, and NewSQL databases, pp. 1–9 (2016)
40. Kepner, J., et al.: Computing on masked data: a high performance method for improving big data veracity, pp. 1–6 (2014)
41. Kepner, J.: Parallel MATLAB for multicore and multinode computers. SIAM (2009)
42. Kepner, J., Cho, K., Claffy, K., Gadepally, V., Michaleas, P., Milechin, L.: Hypersparse neural network analysis of large-scale internet traffic. IEEE (2019). https://doi.org/10.1109/hpec.2019.8916263
43. Kepner, J., et al.: 75,000,000,000 streaming inserts/second using hierarchical hypersparse graphblas matrices (2020)
44. Kepner, J., Jananthan, H.: Mathematics of Big Data: Spreadsheets, Databases, Matrices, and Graphs. MIT Press, Massachusetts (2018)
45. Khan, Y., Zimmermann, A., Jha, A., Gadepally, V., D'Aquin, M., Sahay, R.: One size does not fit all: Querying web polystores. IEEE Access **7**, 9598–9617 (2019)
46. Kim, T., Wang, X., Zeldovich, N., Kaashoek, M.F.: Intrusion recovery using selective re-execution. In: Proceedings of the 9th USENIX Conference on Operating Systems Design and Implementation, OSDI 2010, pp. 89–104. USENIX Association, USA (2010)
47. Klimovic, A., Litz, H., Kozyrakis, C.: Reflex: Remote flash = local flash. In: Proceedings of the Twenty-Second International Conference on Architectural Support for Programming Languages and Operating Systems, ASPLOS 2017, pp. 345–359. Association for Computing Machinery, New York (2017). https://doi.org/10.1145/3037697.3037732
48. Klimovic, A., Wang, Y., Stuedi, P., Trivedi, A., Pfefferle, J., Kozyrakis, C.: Pocket: elastic ephemeral storage for serverless analytics. In: 13th USENIX Symposium on Operating Systems Design and Implementation (OSDI 2018), pp. 427–444. USENIX Association, Carlsbad (2018). https://www.usenix.org/conference/osdi18/presentation/klimovic
49. Larabel, M.: The performance cost to selinux on fedora 31 (2020). https://www.phoronix.com/scan.php?page=article&item=fedora-31-selinux&num=1
50. Leiserson, C.E., et al.: There's plenty of room at the top: what will drive computer performance after Moore's law? Science **368**(6495) (2020)
51. Scaling in the Linux networking stack. https://www.kernel.org/doc/html/latest/networking/scaling.html
52. Lozi, J.P., Lepers, B., Funston, J., Gaud, F., Quéma, V., Fedorova, A.: The Linux scheduler: a decade of wasted cores. In: Proceedings of the Eleventh European Conference on Computer Systems, EuroSys 2016. Association for Computing Machinery, New York (2016). https://doi.org/10.1145/2901318.2901326
53. Lozi, J.P., Lepers, B., Funston, J., Gaud, F., Quéma, V., Fedorova, A.: The Linux scheduler: a decade of wasted cores. In: Proceedings of the Eleventh European Conference on Computer Systems, EuroSys 2016. Association for Computing Machinery, New York (2016). https://doi.org/10.1145/2901318.2901326

54. Lu, J., Holubová, I., Cautis, B.: Multi-model databases and tightly integrated polystores: current practices, comparisons, and open challenges, pp. 2301–2302 (2018)
55. Mao, H., Schwarzkopf, M., Venkatakrishnan, S.B., Meng, Z., Alizadeh, M.: Learning scheduling algorithms for data processing clusters. In: Wu, J., Hall, W. (eds.) Proceedings of the ACM Special Interest Group on Data Communication, SIG-COMM 2019, Beijing, China, 19–23 August 2019, pp. 270–288. ACM (2019). https://doi.org/10.1145/3341302.3342080
56. Mirhoseini, A., Goldie, A., Pham, H., Steiner, B., Le, Q.V., Dean, J.: Hierarchical planning for device placement (2018). https://openreview.net/pdf?id=Hkc-TeZ0W
57. Mitchell, C., Geng, Y., Li, J.: Using one-sided RDMA reads to build a fast, CPU-efficient key-value store. In: Proceedings of the 2013 USENIX Conference on Annual Technical Conference, USENIX ATC 2013, pp. 103–114. USENIX Association, USA (2013)
58. Padioleau, Y., Lawall, J.L., Muller, G.: Understanding collateral evolution in Linux device drivers. SIGOPS Oper. Syst. Rev. **40**(4), 59–71 (2006). https://doi.org/10.1145/1218063.1217942
59. Poddar, R., Boelter, T., Popa, R.A.: Arx: an encrypted database using semantically secure encryption. Proc. VLDB Endowment **12**(11), 1664–1678 (2019)
60. Poimboeuf, J.: Introducing kpatch: dynamic kernel patching (2014). http://rhelblog.redhat.com/2014/02/26/kpatch/
61. Popa, R.A., Redfield, C.M., Zeldovich, N., Balakrishnan, H.: CryptDB: protecting confidentiality with encrypted query processing. In: Proceedings of the Twenty-Third ACM Symposium on Operating Systems Principles, pp. 85–100 (2011)
62. Reuther, A., et al.: Scalable system scheduling for HPC and big data. J. Parallel Distrib. Comput. **111**, 76–92 (2018). https://doi.org/10.1016/j.jpdc.2017.06.009
63. Reuther, A., et al.: Interactive supercomputing on 40,000 cores for machine learning and data analysis. In: 2018 IEEE High Performance extreme Computing Conference (HPEC) (2018). https://doi.org/10.1109/hpec.2018.8547629
64. Shan, Y., Huang, Y., Chen, Y., Zhang, Y.: LegoOS: a disseminated, distributed OS for hardware resource disaggregation. In: Proceedings of the 13th USENIX Conference on Operating Systems Design and Implementation, OSDI 2018, pp. 69–87. USENIX Association, USA (2018)
65. Smalley, S., Vance, C., Salamon, W.: Implementing SELinux as a Linux security module. Technical report (2001)
66. The snowflake cloud data platform. https://www.snowflake.com/
67. Song, C., Lee, B., Lu, K., Harris, W., Kim, T., Lee, W.: Enforcing kernel security invariants with data flow integrity. In: 23rd Annual Network and Distributed System Security Symposium, NDSS 2016, San Diego, California, USA, 21–24 February 2016. The Internet Society (2016). http://wp.internetsociety.org/ndss/wp-content/uploads/sites/25/2017/09/enforcing-kernal-security-invariants-data-flow-integrity.pdf
68. Stonebraker, M., Çetintemel, U.: "One size fits all" an idea whose time has come and gone. In: Making Databases Work: the Pragmatic Wisdom of Michael Stonebraker, pp. 441–462 (2018)
69. Tan, R., Chirkova, R., Gadepally, V., Mattson, T.G.: Enabling query processing across heterogeneous data models: a survey, pp. 3211–3220 (2017)
70. Thumshin, J.: Introduction to the Linux block I/O layer (2016). https://media.ccc.de/v/784-introduction-to-the-linux-block-i-o-layer

71. Tsai, C.C., Jain, B., Abdul, N.A., Porter, D.E.: A study of modern Linux API usage and compatibility: what to support when you're supporting. In: Proceedings of the Eleventh European Conference on Computer Systems, EuroSys 2016. Association for Computing Machinery, New York (2016). https://doi.org/10.1145/2901318.2901341

72. Weisenthal, J.: Reinhart and Rogoff: 'full stop', we made a microsoft excel blunder in our debt study, and it makes a difference (2013). https://www.businessinsider.com/reinhart-and-rogoff-admit-excel-blunder-2013-4

73. Wikipedia: AWS Lambda (2020). https://en.wikipedia.org/wiki/AWS_Lambda

74. Wikipedia: General data protection regulation (2020). https://en.wikipedia.org/wiki/General_Data_Protection_Regulation

75. Attribute-based access control – Wikipedia, the free encyclopedia (2020). https://en.wikipedia.org/w/index.php?title=Attribute-based_access_control&oldid=967477902

76. Completely fair scheduler – Wikipedia, the free encyclopedia (2020). https://en.wikipedia.org/w/index.php?title=Completely_Fair_Scheduler&oldid=959791832

77. Yakoubov, S., Gadepally, V., Schear, N., Shen, E., Yerukhimovich, A.: A survey of cryptographic approaches to securing big-data analytics in the cloud, pp. 1–6 (2014)

78. Zamanian, E., Yu, X., Stonebraker, M., Kraska, T.: Rethinking database high availability with RDMA networks. Proc. VLDB Endow. **12**(11), 1637–1650 (2019). https://doi.org/10.14778/3342263.3342639

79. Zeldovich, N., Boyd-Wickizer, S., Kohler, E., Mazières, D.: Making information flow explicit in HiStar. Commun. ACM **54**(11), 93–101 (2011). https://doi.org/10.1145/2018396.2018419

Polypheny-DB: Towards Bridging the Gap Between Polystores and HTAP Systems

Marco Vogt[(✉)], Nils Hansen, Jan Schönholz, David Lengweiler,
Isabel Geissmann, Sebastian Philipp, Alexander Stiemer, and Heiko Schuldt

Databases and Information Systems Research Group, Department of Mathematics
and Computer Science, University of Basel, Basel, Switzerland
{marco.vogt,nils.hansen,jan.schonholz,david.lengweiler,isabel.geissmann,
sebastian.philipp,alexander.stiemer,heiko.schuldt}@unibas.ch

Abstract. Polystore databases allow to store data in different formats
and data models and offer several query languages. While such polystore
systems are highly beneficial for various analytical workloads, they provide limited support for transactional and for mixed OLTP and OLAP
workloads, the latter in contrast to hybrid transactional and analytical
processing (HTAP) systems. In this paper, we present Polypheny-DB, a
modular polystore that jointly provides support for analytical and transactional workloads including update operations and that thus takes one
step towards bridging the gap between polystore and HTAP systems.

Keywords: Polystore database · Transactional workload · Adaptivity

1 Introduction

In recent years, polystore systems have gained increasing interest in the database
research community. They aim to solve the demand for faster processing of
heterogeneous workloads on massively growing volumes of data.

The idea of a polystore system is to combine polyglot persistence and multistore database systems. Polyglot persistence deals with the challenge of choosing
the right tool (i.e., query language) for a concrete use case: When storing data
used by different types of applications, it might also be beneficial to use different query languages for retrieving the data. Multistore systems, in turn, are
systems which manage data across heterogeneous data stores. All data is accessed
through one interface which supports one query language. A system combining
polyglot persistence and multistore data management is called a *polystore* [13].

Today's IT landscapes usually consist of several applications (e.g., account-
ing, web shop, warehousing, product recommendation engines, newsletters, etc.).
Often, these applications have not been developed in-house but have been
acquired from different software companies. These applications therefore likely
expect different data models and require support for different query languages.

© Springer Nature Switzerland AG 2021
V. Gadepally et al. (Eds.): Poly 2020/DMAH 2020, LNCS 12633, pp. 25–36, 2021.
https://doi.org/10.1007/978-3-030-71055-2_2

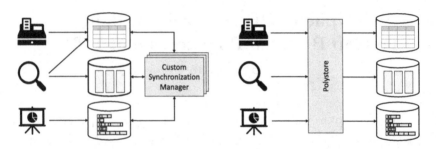

Fig. 1. Replacing a setup consisting of applications directly accessing specialized data storage solutions (left) with a polystore solution deployed between applications and specialized data storage solutions (right).

Often such applications are already shipped with an optimized data storage engine included.

While these optimized data storage engines deliver good performance for the workload of the application they shipped with, this approach only works if the data can be strictly partitioned and separated for the different applications. If this is not the case (which is rather the norm than an exception), at least parts of the data needs to be synchronized using some sort of synchronization approach, as depicted in Fig. 1 (left). Such a setup is not satisfying for multiple reasons: (i) It prevents a close integration of all applications. (ii) The applications cannot run on the latest data. And (iii) there is always the risk of inconsistencies if there is more than one application modifying data.

A possible solution for this scenario is to deploy a polystore system between the applications and the specialized data stores (see Fig. 1, right). Like a federated database system, a polystore allows to access data from different data stores. This ensures a much tighter integration of the application layer and reduces overhead and redundancy in data storage and thus the need for additional synchronization.

In the recent years, different polystore systems have been developed. While they show impressive results in accelerating analytical workloads [4,10], they cannot be used for the scenario depicted in Fig. 1. The main reason is that existing polystores lack support for data manipulation (DML) operations. Since polystore systems are limited to read-only analytical workloads, parallel transactional queries would have to be directly submitted to the underlying data stores, circumventing the polystore layer. This is infeasible, and executing long-running analytical workloads would have a huge impact on the transactional performance of these data storage systems. Hence, a dedicated layer would be needed that is aware of all queries (transactional and analytical); otherwise, it is not possible to prioritize and schedule long-running queries. Furthermore, executing queries on different data stores with parallel workloads can lead to inconsistent data.

Copying transactional data on a regular basis to the polystore solves these problems but does not necessarily allow querying the latest data. This makes it useless for most operational analytical queries requiring the latest data. But also for long-running queries which do not require access to the latest data, this approach causes downtimes of the analytical systems while updating the data. This update process can also massively impact the performance of productive systems. Furthermore, this approach does not offer federated access for transactional workloads.

In parallel to polystores, recently also HTAP (hybrid transactional/analytical processing) systems have gained popularity. These systems provide good performance for mixed transactional and analytical workloads. While typically also offering support for data manipulation, these systems lack common features of a polystore system, especially polyglot persistence and support for a wide variety of workloads since they are usually earmarked for concrete applications and their characteristic workloads.

To deal with mixed, heterogeneous, and dynamically changing workloads (e.g., as they occur in scenarios as the one depicted in Fig. 1) requires a diverse and flexible polystore system that, at the same time, provides the transaction capabilities of an HTAP system. Because there is no off-the-shelf solution for every use case, such an 'HTAP-aware polystore' needs to be flexible in order to allow adding support for a large set of query languages and interfaces as well as underlying data stores.

In this paper we present Polypheny-DB, an implementation of the concepts formerly introduced in [15]. Most importantly, Polypheny-DB goes beyond existing polystores and also adds support for transactional workloads.

The contribution of this paper is twofold: (i) We present the implementation of Polypheny-DB which is to our best knowledge the first polystore system that tries to bridge (or at least narrow) the gap between polystore and HTAP systems and hence jointly offers full support for DML and DDL statements. (ii) We show that the implementation of Polypheny-DB follows a modular approach which can be easily adapted at runtime in various ways (in terms of data stores, data distribution and deployment, and/or query workloads).

The remainder of this paper is structured as follows: Sect. 2 surveys related work. Section 3 introduces Polypheny-DB, Sect. 4 focuses on the flexibility and adaptivity of the system, thanks to its modular architecture, and Sect. 5 concludes.

2   Related Work

The last years have seen a vast proliferation of different types of polystore systems [13]. Every system has a different focus and is developed with a different use case in mind. But all these systems have in common that they provide access to data stored in heterogeneous data stores. In what follows, we introduce and compare selected systems regarding (i) their support for DML and DDL queries, (ii) their architecture in terms of modularity, and (iii) their potential for runtime adaptation.

BigDAWG is one of the pioneer polystore systems. It organizes heterogeneous data stores into "islands" (e.g., relational or array islands) [4]. Every island has an associated query language and data model and connects to one or more data stores. BigDAWG only supports cross-data store queries within the same island. Inter-island queries are not possible without migrating the data first.

The BigDAWG system delivers great results [11] for heterogeneous read-only workloads. However, we did not find any information regarding support for DML or DDL queries. Hence, data needs to be imported into the underlying data stores prior to the start of the system.

BigDAWG's architecture is expandable and allows new data stores and islands to be added, but due to the fact that this requires changes to the Big-DAWG source files, this is not "plug-and-play".

CloudMdsQL is a multistore system [9] that uses an SQL-like query language for querying heterogeneous data stores. It allows subqueries to contain embedded invocations to each data store's native query interface. The authors present an implementation which translates queries into query execution plans and a query engine based on Apache Derby for executing these plans. The focus of their work is on the query language which allows exploiting the full capabilities of the underlying data stores.

The CloudMdsQL system is read-only and has no support for DDL operations. Because only the compiler which parses a CloudMdsQL query and generates the optimized query execution plan is available for download, we cannot discuss the modularity and runtime adaptability of the overall CloudMdsQL system.

Apache Drill [7] is a distributed query engine for interactive analysis of large-scale datasets. The query engine accepts ANSI SQL and MongoDB QL, and supports a variety of NoSQL systems like HBase, Hive, and MongoDB. It also offers support for using relational databases as underlying data stores.

Drill is developed with a focus on extensibility and offers support for pluggable query languages, query planners and query optimizers. It allows new data sources to be added at runtime. Designed as a system targeted on data analytics, it does not support DML queries and its DDL support is limited to linking new data sources.

Hybrid Transactional and Analytical Processing. (HTAP) systems maintain the same data in different storage formats. By leveraging multiple query processing engines, these systems are able to improve performance of read-only workloads [5]. This allows to efficiently perform real-time analytics and transactional workloads on data stored in the same system. Examples for such systems are SAP HANA [3], HyPer [8], and Hyrise [6]. While these HTAP systems are able to provide excellent performance for mixed OLTP and OLAP workloads, they do not qualify as polystores because they do only support one query language.

SnappyData [12] aims to combine the benefits of HTAP systems with the ones of a big data analytics engine. SnappyData uses GemFire as an in-memory transactional data store and combines it with Apache Spark. Having Apache Spark as foundation makes SnappyData very flexible and adds support for a wide range of data stores. It also comes with support for DML and DDL statements. But in contrast to Polypheny-DB, SnappyData treats data stores differently. Data sources in SnappyData are primarily used as source to load data from. While queries on these data sources can be executed in general, this support is limited to the capabilities of the underlying data stores. SnappyData neither allows to replicate data to different data stores nor to partition data.

3   Polypheny-DB

Most polystores are designed with the concept of federated database systems in mind. They enable access to heterogeneous data stored on independent data stores using different data models. The data is usually stored on exactly one of these underlying data stores. As outlined in Sect. 2, these systems are optimized for long-running, read-only queries and offer at most very limited support for DML queries. They furthermore lack support for other typical features of a database management system like transactional execution guarantees.

The focus of existing polystores on long-running queries is fully justified because this is the field where a polystore can massively accelerate query performance as shown in [4,10,14]. Due to the nature of a polystore, it is hardly possible to accelerate simple transactional queries. The challenge is here to keep the overhead as small as possible while still focusing on read-only accesses.

While HTAP systems provide good performance for mixed OLTP and OLAP workload, they miss other important aspects of polystore systems required for the use case depicted in Fig. 1, especially the polyglot persistence. Moreover, in practice, more types of workloads than transactional and analytical can be found and have to be dealt with (e.g., multimedia retrieval or expert systems). Handling these requires the multi-engine and multi-model approach of a polystore system.

We therefore see the demand for a polystore system bridging the gap between HTAP systems and existing polystore systems. Such a system needs to accept queries expressed in the query languages and through the query interfaces demanded by the applications and must be able to efficiently deal with all kinds of workloads generated by the applications. As outlined before, it is furthermore crucial to jointly provide support for DML queries and transactions.

While SnappyData [12] presents an interesting approach to integrate polystore aspects into an HTAP system by combining GemFire with external data sources, with Polypheny-DB we like to introduce an alternative approach. In contrast to SnappyData, Polypheny-DB aims to fully exploit and combine the optimization provided by the underlying data stores. By replicating data to multiple data stores, Polypheny-DB is able to optimize for different types of workloads.

With Polypheny-DB we try to fill the gap between polystore systems and HTAP systems and present a system with a high degree of modularity and

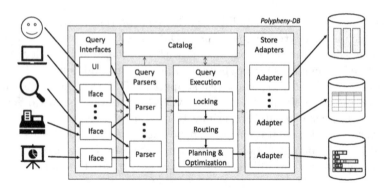

Fig. 2. Simplified architecture of Polypheny-DB. The arrows indicate the execution paths queries can take; the dashed arrows indicate internal communication paths between the modules.

support for adaptions at runtime. In this section, we provide an overview of Polypheny-DB's architecture and design decisions.

3.1 Architecture

Polypheny-DB has been developed with a strong focus on modularity and runtime adaptiveness. Figure 2 depicts the architecture of Polypheny-DB.

Queries are accepted through multiple *query interfaces* supporting one or multiple query languages. A query received through a query interface is forwarded to the matching *query parser*. The parser translates the query into a *logical query plan*. This plan is represented as a tree of relational operators based on an extended relational algebra. This algebra serves as a unified query language and allows to express all query features currently supported by Polypheny-DB.

This logical query plan is then processed by the *locking* module to guarantee the isolation of transactions. When all locks are acquired, the query is processed by the *query router*. This module decides on which of the underlying data stores a query should be executed. This is required because data can be replicated to multiple data stores.

The *routed query plan* is then optimized and translated into a *physical query plan*. This physical query plan contains implementations of the relational operators. Finally, the *data store adapters* take care of translating the physical query plan into the native query language of the underlying data stores. There can be multiple adapters involved in executing a query.

3.2 Schema Management

In Polypheny-DB, we distinguish between two types of schemas: (i) The *logical schema* which represents the structure available to the user, and (ii) the *physical schema* which is maintained on the underlying data stores. The logical schema is expressed and maintained in a relational data model.

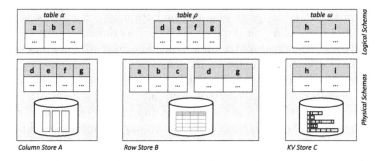

Fig. 3. The columns in tables α and ω only have one placement on the underlying data stores. Table ρ is partially stored on two data stores. All columns of ρ are placed on data store A whereas the columns d and g are additionally stored on data store B.

A major difference of Polypheny-DB compared to other polystore systems is that data stores are not treated as static data sources for a specific dataset but instead as execution engines where data can be placed on. Individual columns or whole tables can be replicated and stored on multiple data stores at the same time. This combination of data replication and vertical partitioning allows to optimize for different types of queries accessing the same data while at the same time minimizing the overhead for DML queries and economically use the available memory and disk space. In Polypheny-DB, the physically stored data is called *placement*. The placements are maintained at the column level. New placements can be added and removed at runtime. Every column needs to have at least one placement. Figure 3 depicts a possible configuration of Polypheny-DB.

A logical schema therefore needs to be mapped to different physical schemas at the same time. The user always interacts with the logical schema, independent of the physical schema(s) the data is represented in. This is especially important when it comes to schema changes which are supported by Polypheny-DB as well.

Another major difference to most existing polystore approaches is the support for schema modifications. The logical schema can be modified at runtime through every query interface supporting schema changes. Currently, the SQL interface and the browser-based user interface come with support for DDL operations. The schema is persistently stored and maintained in the internal catalog. This allows the system to be restarted and all persistent placements to be restored.

3.3 Implementation

Polypheny-DB is implemented in Java. The source is available under the Apache open source license on Github[1]. Major parts of Polypheny-DBs query processing engine are based on Apache Calcite [1]. This allows Polypheny-DB to make use of a reliable cost-based query optimizer also used in several other projects.

---

[1] https://github.com/polypheny/Polypheny-DB.

Query Execution. In the query planning and optimization process, the query is translated into a tree of physical operators. These operators are later translated into the native query language of the underlying data store(s). For every operator (except for the scan operator) exists an implementation which allows every query to be executed purely within Polypheny-DB.

The scan operator—which allows to sequentially read all data from a specified entity—is provided by the data store adapters. In order to make use of the specific optimizations of the underlying data stores, every adapter can additionally provide implementations of the other physical operators.

Polypheny-DB tries to "push down" as many operators to the underlying data stores as possible by using the provided operators. However, if parts of the query plan cannot be "pushed down" (e.g., joining data from different data stores), Polypheny-DB relies on its own implementations. Furthermore, this approach allows to provide support for all query features independent of the data stores on which the data is physically stored.

Transactions. Due to the fact that Polypheny-DB supports DML queries, it requires locking for ensuring the consistency of the data and the correctness of concurrent queries. For the isolation of concurrent queries, we use strong strict two-phase locking [2]. Further, Polypheny-DB provides full transaction support, but only if the underlying data stores offer support for transactions. The system can therefore provide durability—and together with the other guarantees full ACID support—if there is at least one placement for all involved data on a persistent data store.

Data Types. Polypheny-DB supports different integer, float, date, and character types. Within Polypheny-DB, the data is represented and processed using Java types. The mapping to the native data types of the underlying data stores is defined individually for every store. Additionally, Polypheny-DB supports arrays of all supported data types with an arbitrary cardinality and dimension.

4 The Modular Polystore

In this section we particularly focus on the modularity of Polypheny-DB's architecture and its support for DML and DDL statements.

4.1 Query Interfaces

Polypheny-DB can be queried through multiple query interfaces. With the JDBC-SQL interface, we provide an industry standard solution for querying the system using SQL. It offers support for retrieving meta data and for controlling transactions. SQL is the most mature query language supported by Polypheny-DB providing all available query features. It can also be used to manage data stores and other system configurations and the schemas using DDL queries.

With Polypheny-UI, the system has a user interface for managing the schema, modifying the system's configuration, monitoring the status, browsing and modifying the data, managing the underlying data stores, and querying the system by using different query methods and languages.

Polypheny-DB applies the idea of polyglot persistence not only to the data access but also to the management of the system itself. This is achieved by providing two methods for modifying the schema and the system configuration.

Beside JDBC-SQL and the user interface, Polypheny-DB provides a REST-based query interface for accessing and modifying data. Polypheny-DB is also able to directly express queries using its relational algebra representation. This is either possible by submitting the query plan as JSON or by using the graphical builder provided in the Polypheny-UI.

With an Explore-by-Example interface and the dynamic query builder, Polypheny-DB also supports innovative query methods. Explore-by-Example allows to select rows of a result set which should or should not be part of the final result. Polypheny-DB then derives a query fulfilling these requirements. The dynamic query builder assists users in formulating queries containing joins and filters. This is done by dynamically generating a user interface containing specific controls based on a statistical analysis of the currently stored data.

To adapt the system according to the specific requirements of a concrete use case, query interfaces can easily be added, modified, or removed. Furthermore, the modular architecture simplifies the implementation of new query interfaces.

4.2   Query Routing

Query routing is the process of selecting on which of the underlying data stores a query or parts of it should be executed. As depicted in Fig. 3, tables and columns can be partitioned and replicated to multiple underlying data stores.

DML queries always need to be executed eagerly on *all* underlying data stores holding a placement of the modified data. This includes the insertion of new rows. As a consequence, the most suitable placement for executing read-only queries can be arbitrarily selected.

Polypheny-DB currently supports two query routing implementations. The first implementation models the behavior of only storing the data on one of the data stores. The data store is selected when creating the table. This implementation requires no sophisticated query analysis to select the best store and is therefore a suitable base line to compare with the second implementation.

This second implementation is based on the approach introduced in [14]. The basic assumption of this approach is that queries generated by applications are usually derived from templates. Because queries derived from the same template are similar regarding execution time on a specific underlying data store, it is possible to route queries based on structural similarity of queries. While the approach introduced in [14] determines the structural similarity based on the SQL query, the implementation in Polypheny-DB determines the similarity based on the logical query plan.

The query routing is fully modular. New implementations can be provided on classpath without modifying Polypheny-DB. It is also possible to change the router implementation at runtime. It is the responsibility of the query router to make sure that the transition to a different implementation happens without conflicts. This feature is useful if there are maintenance tasks or data imports.

4.3 Data Storage

As a polystore system, Polypheny-DB supports different heterogeneous data stores. These data stores are fully maintained by the system. In order to be able to guarantee correctness, the stores are only accessed through Polypheny-DB.

The specific optimizations and advantages of a data store are exploited by pushing down the entire query or parts of it whenever possible. If supported by the data store, results of a query are only read on demand. Data stores can be added, modified and removed at runtime via the Polypheny-UI or by using SQL commands. The connection to the data stores is handled by means of *adapters*.

An adapter is responsible for everything related to a data store. This includes maintaining the connection and providing the implementations for the supported relational operators. The adapter is also responsible for maintaining the physical schema on the underlying data store and for performing schema migrations. The physical entity names itself are independent from the logical names. An adapter can support multiple database systems.

Polypheny-DB is shipped with adapters for various JDBC-SQL stores including PostgreSQL, MonetDB, and HSQLDB. Furthermore, there is an adapter for Apache Cassandra and for plain CSV files and Polypheny-DB can be extended with new adapters without any modifications due to its strong encapsulation. Some of the provided adapters support an embedded mode to simplify the deployment process.

5 Conclusion and Future Work

With Polypheny-DB, we present a modular polystore system which provides full support for DML and DDL queries. First performance evaluations have shown very promising results which we, due to the lack of space, have not been able to report here. It is planned to present them together with further evaluations we are currently doing and which we plan for our future work. Evaluations of a previous version of our DML-capable query routing system can be found in [14].

Polypheny-DB's modular architecture makes it extensible and adaptable. Several aspects of the system can be modified and exchanged at runtime. In future work, we plan to use this runtime adaptability to automatically adjust the system according to user-defined requirements and changes in the workload. We also plan to expand the set of integrated features. By extending the internal data model, we want to further exploit the advantages of different data models.

Acknowledgments. This work has been partly funded by the Swiss National Science Foundation (SNSF) in the context of the project Polypheny-DB (contract no. 200021_172763).

References

1. Begoli, E., Camacho-Rodríguez, J., Hyde, J., Mior, M.J., Lemire, D.: Apache calcite: a foundational framework for optimized query processing over heterogeneous data sources. In: Proceedings of the 2018 International Conference on Management of Data, pp. 221–230. ACM Press (2018). https://doi.org/10.1145/3183713.3190662
2. Bernstein, P., Hadzilacos, V., Goodman, N.: Concurrency Control and Recovery in Database Systems. Addison-Wesley Longman Publishing Co. Inc., Boston (1987)
3. Färber, F., Cha, S.K., Primsch, J., Bornhövd, C., Sigg, S., Lehner, W.: Sap HANA database: data management for modern business applications. SIGMOD Rec. **40**(4), 45–51 (2012). https://doi.org/10.1145/2094114.2094126
4. Gadepally, V., et al.: The BigDAWG polystore system and architecture. In: 2016 IEEE High Performance Extreme Computing Conference, HPEC 2016, Waltham, MA, USA, 13–15 September 2016, pp. 1–6. IEEE (2016). https://doi.org/10.1109/HPEC.2016.7761636
5. Giceva, J., Sadoghi, M.: Hybrid OLTP and OLAP, pp. 1–8. Springer, Cham (2018). https://doi.org/10.1007/978-3-319-63962-8_179-1
6. Grund, M., Krüger, J., Plattner, H., Zeier, A., Cudre-Mauroux, P., Madden, S.: HYRISE: a main memory hybrid storage engine. Proc. VLDB Endow. **4**(2), 105–116 (2010). https://doi.org/10.14778/1921071.1921077
7. Hausenblas, M., Nadeau, J.: Apache drill: interactive ad-hoc analysis at scale. Big Data **1**, 100–104 (2013). https://doi.org/10.1089/big.2013.0011
8. Kemper, A., Neumann, T.: HyPer: a hybrid OLTP&OLAP main memory database system based on virtual memory snapshots. In: Abiteboul, S., Böhm, K., Koch, C., Tan, K. (eds.) Proceedings of the 27th International Conference on Data Engineering, ICDE 2011, Hannover, Germany, 11–16 April 2011, pp. 195–206. IEEE Computer Society (2011). https://doi.org/10.1109/ICDE.2011.5767867
9. Kolev, B., Bondiombouy, C., Valduriez, P., Jimenez-Peris, R., Pau, R., Pereira, J.: The CloudMdsQL multistore system. In: Proceedings of the 2016 International Conference on Management of Data, SIGMOD 2016, pp. 2113–2116. Association for Computing Machinery, New York (2016). https://doi.org/10.1145/2882903.2899400
10. Kolev, B., Pau, R., Levchenko, O., Valduriez, P., Jimenez-Peris, R., Pereira, J.: Benchmarking polystores: the CloudMdsQL experience. In: 2016 IEEE International Conference on Big Data (Big Data), pp. 2574–2579. IEEE (2016). https://doi.org/10.1109/BigData.2016.7840899
11. Mattson, T., Gadepally, V., She, Z., Dziedzic, A., Parkhurst, J.: Demonstrating the BigDAWG polystore system for ocean metagenomic analysis. In: Proceedings of the 8th Biennial Conference on Innovative Data Systems Research (CIDR), p. 9. http://cidrdb.org/cidr2017/papers/p120-mattson-cidr17.pdf
12. Ramnarayan, J., et al.: SnappyData: a hybrid transactional analytical store built on spark. In: Proceedings of the 2016 International Conference on Management of Data, pp. 2153–2156. ACM Press (2016). https://doi.org/10.1145/2882903.2899408

13. Tan, R., Chirkova, R., Gadepally, V., Mattson, T.G.: Enabling query processing across heterogeneous data models: a survey. In: Nie, J., et al. (eds.) 2017 IEEE International Conference on Big Data, BigData 2017, Boston, MA, USA, 11–14 December 2017, pp. 3211–3220. IEEE Computer Society (2017). https://doi.org/10.1109/BigData.2017.8258302

14. Vogt, M., Stiemer, A., Schuldt, H.: Icarus: towards a multistore database system. In: Proceedings of the 2017 IEEE International Conference on Big Data (Big Data), pp. 2490–2499. IEEE (2017). https://doi.org/10.1109/BigData.2017.8258207

15. Vogt, M., Stiemer, A., Schuldt, H.: Polypheny-DB: towards a distributed and self-adaptive polystore. In: 2018 IEEE International Conference on Big Data (Big Data), pp. 3364–3373. IEEE (2018). https://doi.org/10.1109/BigData.2018.8622353

Persona Model Transfer for User Activity Prediction Across Heterogeneous Domains

Takahiro Hara[✉]

Osaka University, 1-5 Yamadaoka, Suita, Osaka 5650871, Japan
hara@ist.osaka-u.ac.jp
http://www-nishio.ist.osaka-u.ac.jp/~hara/

Abstract. In this keynote talk, we present our project on cross-domain digital marketing where we assume totally different service domains such as Web advertisement domain and E-commerce domain. Cross-domain approaches are useful in situations where some domain does not have enough amount data to develop an accurate prediction model on user activities. Our idea is to transfer persona (user) model from one domain which has richer data to the target domain with less data, i.e., it has a worse prediction model. This project is technically very challenging since we assume totally different domains where the users' activities are different. We present some of recent achievements of our project and also talk about our future plans.

Keywords: Digital marketing · Machine learning · Prediction model

1 Project Overview

This project aims to develop cross-domain approaches based on machine learning and big data processing to provide digital marketing services across different services (domains). While there have been a large number of online services, they generally cannot share raw data such as service usage logs and user IDs due to privacy and right issues, and thus, it often happens that service providers cannot provide sufficient personalized services even if they have a large amount of data in total. Our idea to tackle this problem is developing techniques for persona matching across differently domains and that for transferring prediction models from a source domain to a target domain. With this approach, a domain with less data (i.e., historical data on service usage) can reuse a richer model for user activity prediction built in another domain. Our project is technically very challenging since we assume totally different domains where the users' activities are different. We need to catch some hints to predict user activities in one domain from the model constructed in the other domain in which user activities are totally different. To do so, we develop machine learning and big data processing techniques.

Figure 1 shows the overview of research topics in our project. Below, we briefly present the outline of each topic.

© Springer Nature Switzerland AG 2021
V. Gadepally et al. (Eds.): Poly 2020/DMAH 2020, LNCS 12633, pp. 37–41, 2021.
https://doi.org/10.1007/978-3-030-71055-2_3

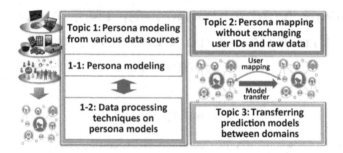

Fig. 1. Overview of our research project

1.1 Topic 1: Persona Modeling from Various Data Sources

Persona modeling is the most fundamental operations in our project. We assume that a persona corresponds to a user, but it can be a virtual user including a representative entity of a group of users. A persona has general attributes such as age, sex, and preferences, and also has activity models, both of which are constructed from data obtained in each domain (e.g., service usage logs). A persona model defines not only each persona but also relationships between personas. It can be represented as a graph where nodes correspond to users and edges corresponds to relationships between users. It also can be represented as embeddings where user vectors represents both the users' characteristics and the relationships with other users simultaneously, e.g., the similarity between two users can be represented as the distance in the vector space.

This topic has a subtopic on data processing on persona models. Because a persona model basically has a very complicated data structure, efficient data processing techniques are essential. Our goal here is to achieve a few decades or hundreds times faster data processing than existing techniques.

1.2 Topic 2: Persona Mapping Without Exchanging User IDs and Raw Data

Identifying same or similar users (user groups) between different domains is useful for effective digital marketing services, e.g. customer transfer across domains. This topic aims to develop matching techniques of same or similar users between domains. Here, we assume that we have bridge users who give us a permission to use their IDs (i.e., ID matching can be made) and their service usage logs in both domains. Therefore, we use data obtained from the bridge users to learn attributive and structural similarity of same or similar users in both domains (in the training phase), and use the findings to identify same or similar (non-bridge) users (in the test phase).

1.3 Topic 3: Transferring Prediction Models Between Domains

This topic addresses the main issue of our project. We assume that there is some latent space which covers all domains. Thus, transferring a prediction model from one (source) domain to another (target) domain is identical to a task of finding a reverse projection function from the source domain to the latent space and then finding a projection function from the latent space to the target domain.

2 Cross-Domain Digital Marketing: Web Advertisement × E-Commerce

We have conducted our first study on cross-domain digital marketing since 2019, where we obtained data from real services in a Web advertisement domain and an E-commerce domain [2]. We performed persona modeling (topic 1), persona matching (topic 2) and cross-domain recommendation (topic 3) as shown in Fig. 2.

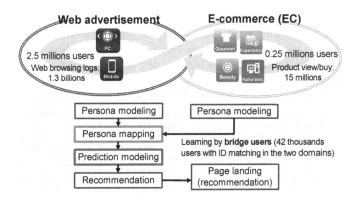

Fig. 2. Web advertisement domain × E-commerce domain

2.1 Persona Modeling

We have developed two different approaches for persona modeling as below.

Content Based Approach. The first one is a word2vec (or content) based approach where documents in Web pages which were browsed by users (i.e., ads were shown on the pages) are used to generate the users' embeddings and documents in product descriptions are used to generate the products' embeddings. For both user and product embeddings, a word2vec technique is used where each dimension in the embeddings corresponds to a same word for both embeddings.

This content based approach aims to tackle to a cold-start problem for both users and products in the e-commerce domain. In most e-commerce services, products on sale quickly change (e.g., in almost every 2 weeks) and most users registered to the services have only a few times or no purchase experiences, i.e., most products and users have no interactions. Thus, it is difficult or almost impossible to effectively model new products and new users from the purchase history (i.e., interactions in the e-commerce domain) and predict such users' purchase activities.

Our idea to solve this cold-start problem is utilizing a cross-domain approach. More specifically, since most people often browse Web pages (i.e., they have enough historical data to model themselves), we try to transfer a rich persona model constructed in the Web advertisement domain to the E-commerce domain. Our hypothesis here is that while these two domains have totally different characteristics, there are some hints in Web browsing pattern to predict user (purchase) activities in the E-commerce domain.

Meta-pass Based Approach. The second approach is a meta-pass (interaction) based approach where the information on interactions between users and Web pages (i.e. Web browsing) and that between users and products (i.e. purchase) are used for persona modeling. The basic idea of using rich information in the Web advertisement domain is the same as that of the first approach, but it is totally different because it does not use any texts in persona modeling.

This approach is motivated by the fact that the first approach (i.e. content based approach) suffers from information losses on user modeling which are caused by blocked accesses, missing links, and meaningless contents. Since the second approach does not use any textual information on user modeling, it does not suffer from such information loses. In addition, the meta-pass based approach has another advantage that it can distinguished two cases in which users browse similar Web pages such as Yahoo! news and Google news. This is because even if two Web pages are similar, these have different URLs. In many cases, such differences in choice of services well represent differences in user preferences.

On the other hand, except for information loss cases, contents generally have richer information than interaction data. Therefore, in total, it depends on situations whether the content based approach or the meta-pass based approach works well.

2.2   Cross-Domain Product Recommendation

After generating user and product embeddings, we apply DMF [3] and NeuMF [1] methods (the original methods have been extended to fit to our problem) to build a prediction model of user purchase activities.

Figure 3 shows a result of performance studies. We compare the top-k hit ratios of our word2vec based methods (denoted by DMF and NeuMF) with some comparison methods including random recommendation and cosine-similarity based method. As a result, we found that our methods significantly outperform the comparison methods. In particular, NeuMF achieved about 26% hit ratio by recommending 10 products (i.e. $k = 10$) among 1500 products on sale during the test period, which is surprisingly high.

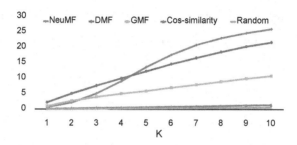

Fig. 3. Performance study

3 Future Plans

We have just started to work on cross-domain approaches for user activity prediction using other domains' data such as public WiFi. We have also worked on environmental modeling using SNS data to catch the trend and user preferences, which can be used as a bias for user activity prediction.

We also plan to investigate the impact of unusual situations such as COVID-19 issue on user activities. It is obvious that user activities significantly changed in such situations, however it is not easy to know how user activity prediction models can adjust to the changes. Therefore, we will work on research of prediction model transfer from ordinary situations to unusual situations.

Acknowledgments. This work was partially supported by JST CREST under Grant J181401085.

References

1. He, X., Liao, L., Zhang, H., Nie, L., Hu, X., Chua, T.-S.: Neural collaborative filtering. In: Proceedings of the International Conference on World Wide Web, pp. 173–182 (2017)
2. Wang, H., et al.: A DNN-based cross-domain recommender system for alleviating cold-start problem. IEEE Open J. Ind. Electron. Soc. **1**, 194–206 (2020)
3. Xue, H.-J., Dai, X.-Y., Zhang, J., Huang, S., Chen, J.: Deep matrix factorization models for recommender systems. In: Proceedings of the International Joint Conference on Artificial Intelligence, pp. 3203–3209 (2017)

PolyMigrate: Dynamic Schema Evolution and Data Migration in a Distributed Polystore

Alexander Stiemer[1(✉)], Marco Vogt[1], Heiko Schuldt[1], and Uta Störl[2]

[1] Databases and Information Systems Research Group, University of Basel,
Basel, Switzerland
{alexander.stiemer,marco.vogt,heiko.schuldt}@unibas.ch
[2] Department of Computer Science, Darmstadt University of Applied Sciences,
Darmstadt, Germany
uta.stoerl@h-da.de

Abstract. In the last years, polystore databases have been proposed to cope with the challenges stemming from increasingly dynamic and heterogeneous workloads. A polystore database provides a logical schema to the application, but materializes data in different data stores, different data models, and different physical schemas. When the access pattern to data changes, the polystore can decide to migrate data from one store to the other or from one data model to another. This necessitates a schema evolution in one or several data stores and the subsequent migration of data. Similarly, when applications change, the global schema might have to be changed as well, with similar consequences on local data stores in terms of schema evolution and data migration. However, the aspect of schema evolution in a polystore database has so far largely been neglected. In this paper, we present the challenges imposed by schema evolution and data migration in Polypheny-DB, a distributed polystore database. With our work-in-progress approach called *PolyMigrate*, we show how schema evolution and data migration affect the different layers of a distributed polystore and we identify different approaches to effectively and efficiently propagate these changes to the underlying stores.

Keywords: Polystore databases · Schema evolution · Data migration

1 Introduction

For several decades, relational databases had a monopoly in the data management layer of information systems. This has changed in the course of the 2000s with the proliferation of novel types of applications, data, and access patterns such as analytical processing on structured data or social graphs [17]. As a consequence, a large variety of different data stores has been introduced, from column stores over key-value stores to document and graph databases. As long as applications are based on rather homogeneous data sets and workloads, these systems

© Springer Nature Switzerland AG 2021
V. Gadepally et al. (Eds.): Poly 2020/DMAH 2020, LNCS 12633, pp. 42–53, 2021.
https://doi.org/10.1007/978-3-030-71055-2_4

are well suited. However, in cases of highly heterogeneous data and/or large fluctuations in the access patterns of applications, not even these specialized systems would be able to provide optimal support.

For this reason, the last years have seen the advent of polystore databases [18], which combine several different, heterogeneous data stores underneath a joint interface. Depending on the type of data to be managed or the access patterns, they may decide to deploy several different data stores and distribute (possibly also partly replicate) data among these stores.

In [20], we have introduced Polypheny-DB, a novel distributed polystore database. Polypheny-DB considers distribution at two levels: At *global level*, data is fragmented and replicated (the latter to increase availability) in order to allocate it to different sites in a global network. Hence, fragmentation, replication, and allocation aims at bringing such data items together that are frequently accessed jointly. The allocation then has to guarantee that the fragments are placed close to their corresponding applications in order to minimize access latency. At *local level*, each site runs an independent polystore that can decide unilaterally, based on a local cost model, which data stores to provide, how to distribute data across these data stores, and how to process queries.

Assume, as an example for an organization running such a distributed polystore, an international auction house with databases and compute centers distributed around the globe (see [20] for more details). The auction house has to jointly deal with several workloads such as Online Transaction Processing (OLTP) (for the actual auctions), Online Analytical Processing (OLAP) (for analyzes of past auctions), graph queries (for recommendations to their customers), and finally also multimedia similarity search queries (to find items and thus auctions based on the visual appearance of the former).

While polystores usually assume the database schema to be static, this is not always the case in practice. Schema evolution—one of the "top ten fears" about the future of databases, according to Stonebraker [16]—needs to be taken into account also in polystores. In the auction house example, changes in the product recommendation engine or additional/revised legal requirements may lead to changes of the logical database schema ("external" reasons, from the polystore's perspective). Because shutting down the entire business of the auction house is not an option, these schema changes and the subsequent data migrations have to be performed efficiently online, without any downtime. In addition, there are also schema changes caused by internal data reorganisation in the local polystores, for instance based on workload changes ("internal" reasons).

In this paper, we introduce *PolyMigrate*, a work-in-progress extension to Polypheny-DB that considers schema evolution and data migration derived from schema changes in a distributed polystore. In particular, we analyze the different layers on which schema changes can originate and how they need to be propagated downwards through the different layers of the polystore (or why and when they do not need to be propagated). To the best of our knowledge, this is the first attempt to address schema evolution and data migration in a polystore context.

The contribution of this paper is threefold: (i) We identify *why* and *where* in the Polypheny-DB stack schema changes might occur and *what* changes need to be considered. (ii) We discuss several options as to *when* these changes have to be propagated downwards in the Polypheny-DB stack to the local data stores, thereby taking into account that potentially large volumes of data might have to be migrated. (iii) We analyze *how* the propagation of schema changes and the migration of data have to be implemented.

The remainder is structured as follows: In Sect. 2 we review related work. Schema evolution and data migration in Polypheny-DB are discussed in Sect. 3. Section 4 presents examples and Sect. 5 concludes.

2 Background and Related Work

In what follows, we introduce basic notions and concepts from database schema evolution, data migration, and multi- and polystores, and survey related work.

2.1 Schema Evolution and Data Migration

Schema evolution in databases is a long investigated and still current topic. In schema-flexible database systems (e.g., NoSQL database systems), schema changes do not have to be executed immediately, but can be divided into two separate steps: *schema evolution* and *data migration*.

Schema evolution describes the changes to the schema without immediately executing them on the data. In this paper, we will discuss *why* and *where* (Sect. 3.1) schema evolution in a distributed polystore is triggered. We also describe *what* types of schema evolution operations occur in Polypheny-DB. We use the operations introduced in [14] and extended for multi-model data in [7]. There are operations on single schema objects (e.g., `add`, `rename`, and `delete`) and operations on multiple schema objects (e.g., `copy`, `move`, `split`, and `merge`).

Data migration follows the schema evolution and is traditionally carried out eagerly, upgrading all legacy data. Yet, in the context of Cloud-hosted data backends, eager migration can be rather costly. Thus, lazy migration may be more cost-efficient, as legacy data objects are only migrated on-the-fly in case they are actually accessed by the application. The downside is that lazy migration introduces a runtime overhead on reads and writes [12]. A compromise between the two competing goals of minimizing latency and migration costs can be reached by migrating hot data predictively [6]. The possible applications of these techniques in Polypheny-DB are discussed in Sect. 3.2.

Related Work. The co-evolution of schemas and the associated XML documents is addressed in [4] and [9], for example. Also in relational databases the handling of different versions in single database systems becomes more and more important [1,5,15]. The schema flexibility of NoSQL database systems brings new challenges for the schema evolution, which are analyzed in [2,12,14] for different types of NoSQL database systems. However, all these approaches only discuss schema evolution in the context of single store databases. Polystore databases lead to new challenges for schema evolution and data migration. In [7], the challenges for multi-model data [8] are outlined in a vision paper. In this paper, we will discuss these challenges and solutions in a more concrete and detailed way using Polypheny-DB.

2.2   Polystore Databases

To efficiently deal with heterogeneous workloads produced by today's zoo of applications, a new generation of database systems has been developed. In the following, we use the taxonomy introduced in [18] to discuss multi- and polystore databases.

When data (or parts thereof) needs to be accessed by different applications, some query languages might be better suited than others—depending on the concrete requirements of the respective applications. *Polyglot persistence* [11,13] addresses this problem and aims at choosing the best suited query language for a concrete use case. This goes back to the concept of *polyglot programming*.

A *polyglot database system* uses a set of *homogeneous* data stores and exposes multiple query interfaces and languages [18]. A *multistore database system*, in contrast, manages data in *heterogeneous* data stores, but offers a common query interface to the outside, as well as only one query language. *Polystore databases* combine the advantages of both polyglot and multistore systems.

Related Work. An example for a multistore system is *Icarus* [19]. Compared to other multistore systems, Icarus always stores all data on all underlying stores and executes incoming queries on the store with the best characteristics for this type of query. Since all data is stored on all stores and Icarus only supports data stores with an SQL interface, there is no need for complex schema and data migrations. *BigDAWG* [3] is a polystore system which organizes heterogeneous data stores into "islands". Each island has an associated query language and data model, (e.g., a relational island is based on the relational data model and exposes an SQL interface). If a query accesses data distributed over different islands, inter-island queries are resolved by migrating data between the islands. *Hybrid.poly* [10] is an in-memory polystore, which is queried using a extended SQL interface. It allows the execution of complex analytical queries on non-relational data being combined with relational data. With *Polypheny-DB* [20] we have introduced a polystore system that does not only provide access to data stored in different kinds of data stores and data models independent of a query language, it can also be deployed in a distributed fashion. Polypheny-DB supports different types of underlying data stores including key-value stores, document stores, plain CSV-files, and relational databases.

3   Schema Evolution and Data Migration in Polypheny-DB

The distributed polystore Polypheny-DB distinguishes two layers (see Fig. 1): At the *global level* ($\mathcal{G}$), data is distributed across several sites in a network, i.e., the global layer consists of interconnected instances. For this, the schema is *fragmented* and *replicated*. The resulting schema fragments are then assigned to sites in the system (*allocation*). At the *local level* ($\mathcal{L}$), each Polypheny-DB site then manages data locally in a polystore ($\mathcal{P}$), i.e., in different data models and data stores. The main objective of the local level is to improve the performance for heterogeneous workloads. There is no communication between the individual data stores which are all considered as black boxes.

Further, applications and clients are not supposed to access the individual data stores directly. Consequently, we follow a top-down approach for data access, which also holds for schema evolution and data migration. Moreover, Polypheny-DB is agnostic to optimizations at the physical level inside the data stores.

To summarize, the global level spans over all instances of the polystore, while the local level only spans over the heterogeneous data stores of a specific polystore instance. Therefore, we distinguish three different types of schemas depicted in Fig. 1: (i) The *global schema* $\mathcal{G}$ (as seen by the applications), (ii) the *local schemas* $\mathcal{L}$ (the schema of an individual polystore instance as a result of fragmentation, replication, and allocation on global level), and (iii) the *physical schemas* $\mathcal{P}$ (the actual schema of an underlying data store).

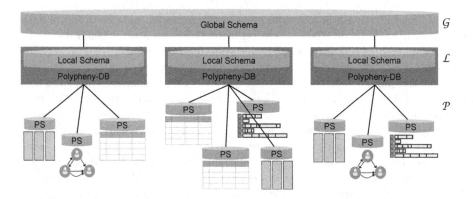

Fig. 1. Different types of schemas in Polypheny-DB. The *global schema* is visible to the application, the *local schema* of an individual polystore instance is the result of the fragmentation and replication process, and the *physical schemas* (PS) are the actual schemas of the data stores. $\mathcal{G}$ denotes the global, $\mathcal{L}$ the local, and $\mathcal{P}$ the physical level.

3.1 Schema Evolution

In the following, we introduce PolyMigrate and analyze the *why*, *where*, *what*, *when*, and *how* regarding schema evolution and data migration in Polypheny-DB.

The "Why": Schema evolution in Polypheny-DB might take place due to one or several of the following three reasons: (i) As a result of an "external" activity like a new version of the application software (e.g., by introducing new features which require additional attributes). (ii) As consequence of an "internal" optimization which leads to a revision of the fragmentation or replication decision at global level. This immediately affects the local schema of at least one of the sites. Internal optimizations might take place due to the attempt to minimize data access latency or request response time. They could also be the result of the re-location of data that is frequently queried together (e.g., by bringing data to the same local site) in order to optimize queries. Other reasons are the minimization of costs (e.g., in the Cloud) or the reduction of the replication degree. And (iii) as a result of the optimization within one polystore instance and the local distribution and allocation of data to one of the underlying data stores.

The "Where": Schema evolution and subsequent migration takes place at each of the three levels in Polypheny-DB. At global level $\mathcal{G}$, schema evolution is triggered by the applications. The fragmentation and replication applied at global level leads to an evolution of one or several of the schemas at local level $\mathcal{L}$. Such inter-model evolution occurs because of global data movements. And finally, the optimizations within a polystore instance (intra-model evolution) lead to the evolution of the individual physical schemas $\mathcal{P}$ of its underlying data stores.

The "What": Schema changes at global level $\mathcal{G}$ are based on the operations introduced in Sect. 2.1 and might thus address only individual schema objects or multiple schema objects jointly. This includes the addition or removal of attributes and alterations of the attribute names or their types. At entity level, the addition of new entities, their deletion, and the split or merge of existing entities need to be considered. Furthermore, changes in the integrity constraints being part of the schema are relevant as well. All these changes at global level need to be properly propagated downwards to the local level.

At local level $\mathcal{L}$, in addition to the changes initialized at global level, schema changes might occur due to a revision of the fragmentation and replication decision and the allocation to sites. At schema level, this means that individual attributes or entire entities will be removed or added from a local site. The allocation is usually determined by the global cost model of Polypheny-DB and is triggered, for instance, by changes in the applications' workloads. Similarly, at the data store level $\mathcal{P}$, aside of schema changes imposed by one of the two layers on top, also internal cost-based optimizations of the local polystore may lead to a re-distribution of entities and a re-allocation of data across its data stores.

3.2   Data Migration

This section discusses aspects related to data migration in PolyMigrate as a result of schema evolution.

The "When": After schema changes have been externally triggered or internally decided, the question is how they are propagated to the next lower level, and how data is handled. For this, five different options can be identified.

First, schema changes and data can be migrated *eagerly*. This means that all the underlying schemas are immediately updated (within the same transaction), and all affected data is migrated to the new schema and/or re-located to the new sites and stores. This guarantees that the entire polystore with all its local sites is up to date at any point in time. However, this comes with significant disadvantages on the performance of the overall system, as all conflicting requests need to be blocked until the migration has successfully completed.

Second, the migration can take place *lazily*. In this case, data is only migrated when accessed. Data which is not requested can still stay in the old schema and at the old site. A request to non-migrated data is temporarily paused and the migration activities for the requested data are triggered. As soon as the migration has succeeded, the request is resumed and the requested data is returned. Therefore, such a lazy migration increases the access latency for the first access to a data item after a schema change, but it does not require a costly eager migration [6,12]. However, in lazy migration, data in the old and the new schema might temporarily co-exist. Even several versions of the revised schema could stay in the system when schema changes appear frequently. It has to be noted that lazy migration is only well suited for OLTP workloads. OLAP queries would trigger a complete migration of the data, resulting in an enormous impact on query latency (and a query would have to be paused for a significant duration).

Third, a compromise between the two competing goals of minimizing latency and migration costs can be reached by a *proactive* strategy. Data is migrated in a background process with the objective that the migration is completed as soon as data is requested. This can be done, for example, by predicting future data accesses based on access statistics and a suitable prediction function [6]. In case data is requested that has not been migrated yet, the lazy approach is applied. Proactive migration is also more suitable for OLTP than for OLAP workloads.

A fourth option is to refrain from physically implementing the schema changes at all. This can be done by using *query rewrites* instead. Any access to a data item (specified in terms of the new schema) will be addressed by a re-written query that exploits the old (and still physically present) schema. While this option has the least effects on the run-time behavior of the system as all changes are not propagated, it is limited to a subset of schema manipulations only (e.g., a `rename` of an attribute, or the `split` or `merge` of an entity). Furthermore, the query rewrite quickly becomes complex if several such changes occur consecutively. In this case, also query re-write would have to be cascaded accordingly.

Fifth, the *incremental* approach combines lazy schema migration and query re-write. The incremental approach propagates schema changes and migrates data when the system's load is below a certain threshold. This ensures that regular operations are not unnecessarily affected by migration activities. For this, data is sequentially migrated, starting from the most recently migrated data item. If the system's load exceeds the threshold, the incremental migration is paused. Access to data that are not yet migrated is subject to a query re-write.

While all these approaches are complementary, they can be seamlessly combined in one system. First, all local polystores are autonomous and can independently implement the schema changes imposed by the global level. One polystore, for example, might decide to perform schema changes eagerly, a second one follows a lazy approach, and a third one relies on query re-writing. Similarly, even within one polystore, several approaches may co-exist (e.g., using a proactive approach for a subset of the data and the incremental approach for the rest).

The "How": Addresses the propagation of schema changes downwards to the local data stores and in particular the handling of data migration.

If schema changes originate at global level $\mathcal{G}$, they are propagated down to the local level $\mathcal{L}$. There, they can be subsequently implemented; in this case, data needs to be migrated accordingly. Alternatively, depending on the nature of the changes, the schema changes are not implemented (and data is not migrated). But this requires the re-write of every query at the local level. Similarly, if the changes are implemented at level $\mathcal{L}$, they will be propagated to the data store level $\mathcal{P}$. There, they can again be enacted, with subsequent data migration, or be avoided using query re-writes.

If schema changes originate at the $\mathcal{L}$ level (i.e., due to a revised fragmentation, replication, or allocation), they will be propagated down to the underlying data stores $\mathcal{P}$. These changes then require data to be migrated once they are implemented in the physical schema, or they are again handled by query re-writes.

Finally, schema changes and data migration activities might start at and only affect one of the data stores at $\mathcal{P}$ level at a site where they will then be enacted.

4   Sample Scenarios and Recommendations

In this section we illustrate the techniques introduced in Sect. 3 with concrete examples from the online auction house scenario mentioned in Sect. 1.

4.1   Global Schema Changes

Archiving: Consider the fictitious international auction house and assume that each completed auction will be archived, for legal and auditing reasons. However, the information that needs to be archived is only a subset of all the information collected on the good which was sold, the buyer and the seller. Therefore, in an attempt to condense the information and to reduce the number of relations per auction while still meeting all legal auditing requirements, a denormalization of the schema at global level $\mathcal{G}$ is applied for archival purposes:

```
COPY     U.last_name, U.first_name, U.zip_code, U.country,
         S.last_name, S.first_name, S.zip_code, S.country,
         B.amount, B.timestamp
TO       A
   WHERE A.id = B.auction  AND  U.id = B.user  AND  S.id = A.user
```

where A is the relation holding the auction, B holds the final bid, U is the buyer relation and S the seller. Since the archive is only highly infrequently accessed (if at all), schema evaluation is propagated in an incremental approach and the local schema at $\mathcal{L}$ is only updated when data is accessed.

Data Protection: Assume that new data protection laws in one country where the auction house does business require a change in the data distribution plan. Currently, sensitive user data is stored at the cheapest Polypheny-DB instance $\mathcal{L}_{cheap}$ which happens to be located in another country. Since data protection regulations require sensitive data not to leave the country, the database administrators of the auction house implement new constraints. At the global level $\mathcal{G}$, Polypheny-DB uses split and merge operations to create new entities which separate sensitive from insensitive data. In general, the propagation of the schema changes follow an incremental approach; however, this approach needs to be switched to eager propagation shortly before the laws come into force.

4.2   Local Schema Changes

Data Protection (continued): Consider again the data protection scenario from Sect. 4.1. After new entities have been created at level $\mathcal{G}$ to separate sensitive from insensitive data (only for the country affected by new legal constraints; for all other countries, the schema is unaltered), the insensitive part can still stay at the $\mathcal{L}_{cheap}$ instance while the sensitive part needs to be migrated to a local instance $\mathcal{L}_{local}$. In analogy to level $\mathcal{G}$ propagation, migration takes place incrementally, with a transition to eager shortly before the laws come into force.

```
MOVE     U.address, U.birthdate, U.credit_rating
TO       SD                                      -- Sensitive data
   WHERE SD.uid = U.id  AND  U.country = 'CH'
```

Replication: In an attempt to minimize access latency, auctions are stored on local servers, close to the location of the seller and the (majority) of the buyers. According to a new legal policy of the Swiss branch of the auction house, auctions exceeding a certain threshold need to be stored redundantly on an instance in Switzerland. Therefore, a materialized view will be created and deployed at $\mathcal{L}_{CH}$. This materialized view replicates auction data subject to the new regulations and leads to an update of the entire data distribution scheme. In order to implement the policy, all auction-related data contained in this materialized view has to be migrated eagerly to $\mathcal{L}_{CH}$, due to legal requirements.

4.3   Physical Level Schema Changes

Specialized Data Store: The local German Polypheny-DB instance $\mathcal{L}_{DE}$ measures an increase in visual similarity searches. So far, the similarity features were stored in the relation IMG in a row store $\mathcal{P}_{RS}$. To speedup the similarity searches, the Polypheny-DB instance decides to deploy a new store $\mathcal{P}_{sim}$ optimized for queries addressing visual similarity. The migration procedure, first, separates the similarity features from the $\mathcal{P}_{RS}$ entity using the split operation. After the creation of the schema in $\mathcal{P}_{sim}$, the schema dealing with the similarity features is moved to $\mathcal{P}_{sim}$. Finally, the data is moved on a proactive approach. The feature relation F is created by using the split operation (note that F will be moved to $\mathcal{P}_{sim}$ while IMG remains on $\mathcal{P}_{RS}$; further, the logical schema on $\mathcal{L}_{DE}$ is unchanged):

```
SPLIT  IMG
INTO   IMG.filename, IMG.type, IMG.size, IMG.auction -- Image data
AND    F.filename, F.features                         -- Feature data
```

Self-optimization: The local polystore instance at site $\mathcal{L}$ detects a change in the local workload. As a consequence of the changed interest of users, attributes are jointly accessed for which the physical schemas are not (yet) prepared since these attributes are assigned to different stores (different physical schemas). Hence, in order to optimize the physical schemas at $\mathcal{L}$, attributes are moved from one store $\mathcal{P}_1$ to another store $\mathcal{P}_2$. This schema change, together with the corresponding data migration, is done in a lazy approach. In case the entire workload further increases significantly, then even a new physical store $\mathcal{P}_{new}$ can be deployed at $\mathcal{L}$ and data be copied incrementally.

5   Conclusions and Outlook

Polystores allow to address heterogeneous and dynamic application workloads by jointly considering several different data stores. Data can then be provided in different models and systems, and when changes to the workload are detected, data can be migrated across stores. In this paper, we have introduced PolyMigrate, a work-in-progress approach that considers various options to apply and propagate schema changes and data migration in the distributed polystore Polypheny-DB. We plan to thoroughly evaluate and compare these alternatives based on the auction house benchmark [20] tailored to polystore databases.

Acknowledgment. This work has been partly funded by the Swiss National Science Foundation (project *Polypheny-DB: Cost- and Workload-aware Adaptive Data Management*, no. 200021_172763) and the German Research Foundation (project *NoSQL Schema Evolution and Big Data Migration at Scale*, no. 385808805).

References

1. Ataei, P., Termehchy, A., Walkingshaw, E.: Variational databases. In: Proceedings of the 16th International Symposium on Database Programming Languages, DBPL 2017, Munich, Germany, 1 September (2017). https://doi.org/10.1145/3122831.3122839
2. Bonifati, A., Furniss, P., Green, A., Harmer, R., Oshurko, E., Voigt, H.: Schema validation and evolution for graph databases. In: Laender, A.H.F., Pernici, B., Lim, E.-P., de Oliveira, J.P.M. (eds.) ER 2019. LNCS, vol. 11788, pp. 448–456. Springer, Cham (2019). https://doi.org/10.1007/978-3-030-33223-5_37
3. Duggan, J., et al.: The BigDAWG polystore system. ACM SIGMOD Rec. **44**(2), 11–16 (2015). https://doi.org/10.1145/2814710.2814713
4. Guerrini, G., Mesiti, M., Sorrenti, M.A.: XML schema evolution: incremental validation and efficient document adaptation. In: Barbosa, D., Bonifati, A., Bellahsène, Z., Hunt, E., Unland, R. (eds.) XSym 2007. LNCS, vol. 4704, pp. 92–106. Springer, Heidelberg (2007). https://doi.org/10.1007/978-3-540-75288-2_8
5. Herrmann, K., Voigt, H., Behrend, A., Rausch, J., Lehner, W.: Living in parallel realities: co-existing schema versions with a bidirectional database evolution language. In: Proceedings of the 2017 ACM International Conference on Management of Data, SIGMOD Conference 2017, Chicago, IL, USA, 14–19 May 2017, pp. 1101–1116. ACM (2017). https://doi.org/10.1145/3035918.3064046
6. Hillenbrand, A., Levchenko, M., Störl, U., Scherzinger, S., Klettke, M.: MigCast: putting a price tag on data model evolution in NoSQL data stores. In: Proceedings of the 2019 International Conference on Management of Data (SIGMOD 2019), pp. 1925–1928. ACM, Amsterdam (2019). https://doi.org/10.1145/3299869.3320223
7. Holubová, I., Klettke, M., Störl, U.: Evolution management of multi-model data. In: Gadepally, V., et al. (eds.) DMAH/Poly -2019. LNCS, vol. 11721, pp. 139–153. Springer, Cham (2019). https://doi.org/10.1007/978-3-030-33752-0_10
8. Lu, J., Holubová, I.: Multi-model databases: a new journey to handle the variety of data. ACM Comput. Surv. **52**(3), 55:1–55:38 (2019). https://doi.org/10.1145/3323214
9. Necaský, M., Klímek, J., Malý, J., Mlýnková, I.: Evolution and change management of XML-based systems. J. Syst. Softw. **85**(3), 683–707 (2012). https://doi.org/10.1016/j.jss.2011.09.038
10. Podkorytov, M., Soderman, D., Gubanov, M.: Hybrid.poly: an interactive large-scale in-memory analytical polystore. In: 2017 IEEE International Conference on Data Mining Workshops (ICDMW), pp. 43–50. IEEE (2017). https://doi.org/10.1109/ICDMW.2017.13, http://ieeexplore.ieee.org/document/8215643/
11. Sadalage, P., Fowler, M.: NoSQL Distilled: A Brief Guide to the Emerging World of Polyglot Persistence. Addison-Wesley Professional, Boston (2012). 0321826620
12. Saur, K., Dumitras, T., Hicks, M.W.: Evolving NoSQL databases without downtime. In: 2016 IEEE International Conference on Software Maintenance and Evolution (ICSME 2016), pp. 166–176. IEEE Computer Society, Raleigh (2016). https://doi.org/10.1109/ICSME.2016.47
13. Schaarschmidt, M., Gessert, F., Ritter, N.: Towards automated polyglot persistence. In: Proceedings of Datenbanksysteme für Business, Technologie und Web (BTW), 16. Fachtagung des GI-Fachbereichs "Datenbanken und Informationssysteme" (DBIS), Hamburg, Germany, 4–6 March 2015. LNI, vol. P-241, pp. 73–82. GI (2015). http://subs.emis.de/LNI/Proceedings/Proceedings241/article46.html

14. Scherzinger, S., Klettke, M., Störl, U.: Managing schema evolution in NoSQL data stores. In: Proceedings of the 14th International Symposium on Database Programming Languages (DBPL 2013), Riva del Garda, Trento, Italy. (2013). http:// arxiv.org/abs/1308.0514

15. Spoth, W., et al.: Adaptive schema databases. In: 8th Biennial Conference on Innovative Data Systems Research, CIDR 2017, Chaminade, CA, USA, 8–11 January 2017 (2017)

16. Stonebraker, M.: My top ten fears about the DBMS field. In: Proceedings of the 34th IEEE International Conference on Data Engineering (ICDE 2018), pp. 24–28. IEEE Computer Society, Paris (2018). https://doi.org/10.1109/ICDE.2018.00012

17. Stonebraker, M., Çetintemel, U.: "One size fits all": an idea whose time has come and gone. In: Brodie, M.L. (ed.) Making Databases Work: the Pragmatic Wisdom of Michael Stonebraker, pp. 441–462. ACM/Morgan & Claypool (2019). https:// doi.org/10.1145/3226595.3226636

18. Tan, R., Chirkova, R., Gadepally, V., Mattson, T.: Enabling query processing across heterogeneous data models: a survey. In: 2017 IEEE International Conference on Big Data (Big Data), pp. 3211–3220. IEEE, Boston (2017). https://doi. org/10.1109/BigData.2017.8258302

19. Vogt, M., Stiemer, A., Schuldt, H.: Icarus: towards a multistore database system. In: Proceedings of the 2017 IEEE International Conference on Big Data (Big Data), pp. 2490–2499. IEEE, Boston (2017). https://doi.org/10.1109/BigData. 2017.8258207

20. Vogt, M., Stiemer, A., Schuldt, H.: Polypheny-DB: towards a distributed and self-adaptive polystore. In: Proceedings of the IEEE International Conference on Big Data (Big Data 2018), pp. 3364–3373. IEEE, Seattle (2018). https://doi.org/10. 1109/BigData.2018.8622353

An Architecture for the Development of Distributed Analytics Based on Polystore Events

Athanasios Zolotas[1]([✉]), Konstantinos Barmpis[1], Fady Medhat[1],
Patrick Neubauer[1], Dimitris Kolovos[1], and Richard F. Paige[1,2]

[1] Department of Computer Science, University of York, York, UK
{thanos.zolotas,konstantinos.barmpis,fady.medhat,patrick.neubauer,
dimitris.kolovos,richard.paige}@york.ac.uk
[2] Department of Computer Science, McMaster University, Hamilton, Canada

Abstract. To balance the requirements for data consistency and availability, organisations increasingly migrate towards hybrid data persistence architectures (called *polystores* throughout this paper) comprising both relational and NoSQL databases. The EC-funded H2020 TYPHON project offers facilities for designing and deploying such polystores, otherwise a complex, technically challenging and error-prone task. In addition, it is nowadays increasingly important for organisations to be able to extract business intelligence by monitoring data stored in polystores. In this paper, we propose a novel approach that facilitates the extraction of analytics in a distributed manner by monitoring *polystore queries* as these arrive for execution. Beyond the analytics architecture, we presented a pre-execution authorisation mechanism. We also report on preliminary scalability evaluation experiments which demonstrate the linear scalability of the proposed architecture.

Keywords: Analytics · Hybrid databases · Polystores · Queries

1 Introduction

Data managed within an organisation may have significantly variable consistency and availability requirements. For example, in the case of an e-commerce system, data used to provide recommendations of products to users needs to be highly available but the consistency of such data is not critical. By contrast, for other subsets of data, such as data recording customer payments, compromising data consistency to improve availability is not acceptable. As a result, organisations increasingly need to use both relational and non-relational databases.

Nowadays, small businesses to big corporations use monitoring tools and data analytics to extract business intelligence based on data stored in such hybrid database systems. This can lead to improvement on their systems and business processes enhancing the customer experience. For example, in an e-commerce system retailers often need to identify relationships of interest between products

© Springer Nature Switzerland AG 2021
V. Gadepally et al. (Eds.): Poly 2020/DMAH 2020, LNCS 12633, pp. 54–65, 2021.
https://doi.org/10.1007/978-3-030-71055-2_5

they trade to provide useful recommendations to customers. Such knowledge can be extracted by including analytics logic within the application business logic to store into the database information of interest.

We propose in this work another approach, that of monitoring the polystore queries and extract analytics of interest as queries arrive for execution. This approach comes with the benefit of calculating analytics in real-time without the need of mixing analytics with core business logic. It also offers access to data that may never be stored in a database (e.g., from "select" queries) or data that were later deleted or updated. Consider the following two motivating scenarios in the domain of an e-commerce website. Analytics developers are able to monitor "trending products" by monitoring the number of "select" queries arriving to the database for each product by the web application when users browse to the details page of each product. In a modern large-scale system not all user requests would end up in the database - given that there are HTTP-level caches commonly in place in such systems. In addition, developers can identify products that users *almost* bought by checking pairs of insert and delete queries to each user's basket for the same product.

In this paper we propose an architecture that consumes queries on polystores to facilitate orthogonal real-time monitoring and predictive analytics. To accommodate the large number of events that polystore-backed applications are expected to generate in real-world scenarios, the proposed architecture is implemented on top of proven big-data-capable frameworks such as Apache Flink [5] and Apache Kafka [4]. Flink is used for distributing the processing/execution workload of analytics applications while Kafka stores and dispatches the generated events in a form of a distributed log. Beyond the possibility of producing analytics based on queries, the analytics component also offers a mechanism of blocking the execution of commands that do not meet developer-defined criteria.

2 Background

The EU-funded Horizon 2020 project TYPHON [6] has developed a model-based methodology and integrated technical offering for designing, developing, querying, evolving, analysing and monitoring scalable hybrid data persistence architectures. It is based on three Domain-Specific Languages (DSLs), namely TyphonML, TyphonDL and TyphonQL which facilitate designing, deploying and querying hybrid data-stores, respectively.

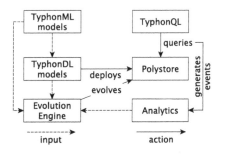

Fig. 1. An overview of the architecture of TYPHON.

Figure 1 shows an overview of the TYPHON architecture and in the following we briefly present the three aforementioned languages. Interested readers can find out about TYPHON and its different components in [6].

TyphonML. TyphonML is a textual modelling language that supports the design of hybrid polystores. Developers, using TyphonML, create models that include information regarding the concepts appearing in the polystore, their fields and their relationships. These models also include information about the databases that are involved in the system. As a result, they represent the high-level infrastructure of a hybrid polystore. An example is shown in Fig. 2a.

```
1⊖ entity Product {                          1  import EcommerceEnhanced.xmi
2      id : string[64]                        2  import RelationalDatabase.tdl
3      name : string[64]                      3  import DocumentDatabase.tdl
4      description : string[1024]             4  import dbTypes.tdl
5      category -> Category[1]                5  containertype Docker
6  }                                          6  clustertype DockerCompose
7                                             7  platformtype localhost
8⊖ relationaldb RelationalDatabase{           8⊖ platform platformName : localhost {
9      tables{                                9⊖     cluster clusterName : DockerCompose {
10⊖        table {                            10⊖        application Polystore {
11            ProductDB : Product             11⊖            container RelationalDatabase : Docker {
12⊖            index productIndex{             12                deploys RelationalDatabase
13                attributes ('Product.name') 13⊖                ports {
14            }                               14                    target = 3306 ;
15            idSpec ('Product.name')         15                }
16        }                                   16            }
17    }                                       17⊖            container DocumentDatabase : Docker {
18  }                                         18                deploys DocumentDatabase
                                              19⊖                ports {
                                              20                    target = 27017 ;
                                              21                }
                                              22            }
                                              23        }
                                              24    }
                                              25  }
```

(a) TyphonML syntax example.   (b) TyphonDL syntax example.

Fig. 2. Example of the TyphonML and TyphonDL syntaxes.

TyphonDL. Arguably, the abstraction gap between high-level TyphonML models and ready-to-use polystores is not negligible. To bridge this gap, an intermediate polystore deployment modelling language (TyphonDL) is used. TyphonML models are transformed to *TyphonDL models* and are enhanced with more fine-grained database-specific configuration details. TyphonDL models represent the deployment infrastructure of that polystore in terms of the specific cloud platform and deployment tools employed and are used to generate the necessary installation and configuration facilities that, when executed, can assemble the polystore in an automated manner. An example is shown in Fig. 2b.

TyphonQL. As data in a TYPHON polystore is distributed across a number of heterogeneous databases a common data manipulation language is used. TyphonQL is developed for performing data manipulation commands (e.g., insert, delete, etc.). Since TyphonQL queries[1] are only executable on polystores precisely specified using TyphonML and TyphonDL, dedicated compilers/interpreters exploit this rich structural and semantical information to type-check and transform TyphonQL queries to high-performance native queries and APIs.

---

[1] An example TyphonQL "select" query: **from** User u **select** u.age **where** u.id == 1.

3   Proposed Architecture

An overview of the developed polystore data event publishing and processing architecture is shown in Fig. 3. Events undergo nine stages that are distributed among two main interleaved phases; authorisation and analytics. The authorisation phase involves validating if a new incoming (TyphonQL) query will be allowed execution by the polystore or not. The rules defined in authorisation tasks can be based either on hardcoded conditions (e.g., value of a specific field is above a threshold) or on information extracted from the history of previous events processed through the stream processor.

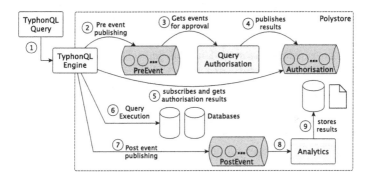

Fig. 3. The proposed data event publishing and processing architecture.

Such a query authorisation mechanism would be also possible at the application layer. An advantage though of using query authorisation at the level of the polystore is that many applications that connect to the same polystore can reuse those authorisation rules instead of having to implement and maintain them individually in each application.

The second phase of analytics involves the continuous consumption of TyphonQL queries that were executed. The tasks developed at this stage consume PostEvent objects (the structure of these objects is described in Sect. 3.1). Below are the stages an event will go through as it progresses within the proposed architecture (numbers in the list correspond to those shown in Fig. 3).

1. A query is passed by a user to the TyphonQL engine for execution.
2. TyphonQL publishes a pre-execution event (PreEvent) for the query, push it to a pre-event queue and waits for an authorisation decision of this event.
3. A stream processor (Apache Flink in the current implementation) dedicated to authorisation, consumes messages from the authorisation queue to apply the configured authorisation checks (presented in Sect. 3.2) before generating an authorisation decision of an event.
4. Following the application of the required authorisation checks, the stream processor publishes the authorisation decision to an authorisation queue.

5. TyphonQL receives the authorisation decision it was waiting for.
6. Based on the outcome of the authorisation decision, TyphonQL will execute the query received at step 1 or reject it.
7. A PostEvent object is generated and pushed to the analytics queue.
8. The analytics stream processor consumes the (post) event to which the relevant analytics tasks (described in Sect. 3.3) could be applied.
9. The results of the analytics can be stored/published using different mechanisms (e.g., a database, a filesystem, a web-service).

3.1 Data Event Structure

This section summarises the data analytics events structure metamodel which is presented in Fig. 4. Both PreEvents and PostEvents have a unique *id* and store the TyphonQL *query* that generated them. The time when the query arrived for execution to the polystore is stored in the *queryTime* attribute of PreEvents. A boolean flag, named *authenticated*, stores the result of the decision on if the query is approved for execution or rejected. The *resultSetNeeded* boolean property declares if the polystore needs to store the result of the execution of the commands in the PostEvent object after it is executed. The *slots* list acts as an extension mechanism to be utilised by polystore-backed applications. It hosts key-value pairs of any custom properties that analytics developers need to pass to the analytics workflow to accommodate their requirements when manipulating the events. A use of this feature is demonstrated in the scenario presented in the evaluation (see Sect. 4).

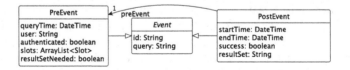

Fig. 4. The event metamodel.

PostEvent instances will point to their corresponding PreEvent instance. PostEvents also hold timestamps of when the execution started (*startTime*) and when it ended (*endTime*). PostEvents store a *success* flag declaring whether the execution of the query was successful or not. Finally, the result set returned from the execution of the command is stored in the *resultSet* attribute.

3.2 Authorisation Tasks

The event authorisation architecture is based on the concept of authorisation tasks. Each task contains logic that decides if a query should be executed or not against the polystore. In the proposed authorisation architecture, all the configured authorisation tasks are part of an authorisation chain. A PreEvent arriving for authorisation, visits authorisation tasks one after the other, unless a previously visited task has already rejected the execution of that event. A query is executed if it has been approved (or ignored) by all the tasks in the chain.

Developers can provide the aforementioned logic by implementing an abstract class (namely *GenericAuthorisationTask*) which is part of the analytics infrastructure. More specifically, they need to implement two methods for each authorisation task: (i) the *checkCondition(Event event)* and (ii) *shouldIReject(Event event)*. The first method (i.e., *checkCondition(...)*), checks if the task is responsible for approving or rejecting a query. The second method (i.e., *shouldIReject(...)*), includes the logic that defines if a query should be approved or rejected. The *shouldIReject* method is called if and only if the *checkCondition* method of the task evaluates to true.

The authorisation chain is built using the Flink's concept of side outputs [11]. Each stream of data in Flink can be transformed to another stream in which the data is grouped using tags based on some logic defined in the transformation operator. The analytics architecture automatically tags PreEvents into specific groups that facilitate the orchestration of the flow of events within the authorisation chain. More specifically, all rejected events, no matter which task rejected them, end up in a group tagged "Rejected". The events that were either approved or not checked (because of the "checkCondition" method returning "false") by a task are placed under the group which is tagged by the name of the Task. Those are given as input to the next task in the chain where the process is repeated.

An orchestrator application is automatically generated (in Java) using a purpose-built code generator. The orchestrator subscribes to the pre-event queue, consumes the event stream and is responsible for re-directing the events to the appropriate tasks based on the results of each task's *checkCondition* and *shouldIReject* methods.

3.3 Analytics Tasks

Analytics tasks are implemented as individual Flink jobs. Each analytics task needs to implement the *analyze* method of an *IAnalyzer* interface provided by the architecture. The *analyze* method automatically subscribes to the post-event queue and provides a Flink datastream of PostEvent objects to the method. Developers are able to define their scenarios by using Flink's the built-in stream operators (e.g., map, process, aggregate, sum, etc.). A provided class that includes the main method for calling the analytics tasks is then used to deploy the scenario in a Flink execution environment.

3.4 Deployment

The analytics and authorisation tasks can be deployed in a Flink/Kafka infrastructure. This can be achieved by using one of the available containerized deployments (i.e., Docker and Kubernetes). TyphonDL (see Sect. 2) generates the necessary deployment scripts. For Docker, we use the wurstmeister Zookeeper[2]

---

[2] https://hub.docker.com/r/wurstmeister/zookeeper/.

and Kafka[3] DockerHub images. The Kubernetes deployment is based on the Strimzi [9] package. Flink cluster deployment is achieved by using the official Apache Flink cluster deployment scripts [10].

4   Scalability Evaluation

The evaluation of the scalability of the proposed architecture requires ingestion of large volumes of data. In order to evaluate our work we developed an e-shop simulator that produce large volumes of synthetic, but realistic, data. The e-shop simulator is based on the notion of "Agents". An agent simulates the behaviour of one type of shopper (i.e., a User) in an e-shop. Developers can use either the *executeQuery(...)* method to execute a query against the polystore or the *createAndPublishPostEvent(...)/createAndPublishPreEvent(...)* to skip the execution of the command against the polystore and create directly a PostEvent/PreEvent object in the relevant analytics queues. To be able to evaluate the scalability of the proposed architecture, we opted for the latter option avoiding the overhead of having to wait for the execution of the actual command against the database in order to produce the Pre/PostEvent object.

4.1   Authorisation Chain Scalability Evaluation

In order to test the scalability of the authorisation chain, we produced an increasing number of events which were given as input into the Pre-Event queue. More specifically, users (agents) were simulating the placement of orders in the e-shop. TyphonQL "insert" commands (see Listing 1.1) were generated for the placed orders including details of the credit card used to pay the order. Three authorisation tasks were created, applying different validation rules on the credit card used. The first task checks the existence of a credit card in the query, the second if the credit card has expired and the third if the credit card number was valid.

Listing 1.1. An example TyphonQL insert command used in the experiment.

```
insert Order {id:'...', date:'...', total:'...', products:[...], user:'...',
paidWith: CreditCard {id:'...', number:'6007-2216-3740-9000',
expiryDate:'2021-06-25T08:36:13.656}}
```

As described in Sect. 3, if an event is rejected by one authorisation task, it is not passed to the following task(s) in the chain but is directed automatically to the authorisation queue as rejected. In the simulator, the agents were producing orders that had always a credit card assigned to them, so they were approved by the first task. From those, half (50%) were having an expired credit card attached to them thus, they were rejected from the second task. Those passed successfully from the second task have a 50% chance of having an invalid credit card number. Following this pattern, we increased the variability as some of the events will be passing the whole chain, while some will be rejected earlier.

---

[3] https://hub.docker.com/r/wurstmeister/kafka/.

The chain was deployed in a cluster of three machines; one acting as the master and the other two as the workers[4]. In our experiment, the master was also hosting the relevant Kafka topics (PRE and AUTH). We restricted the Flink deployment to allocate and use only 8 GB of the available 64 GB for each worker.

Figure 5 shows the total execution time (in seconds) for processing all the event and posting the (rejected/approved) PreEvent in the authorisation queue. The graph shows linear scalability which confirms our expectation as the analytics architecture is built atop tools such as Apache Flink and does not add any bottleneck. The master node's average memory increases steadily and averages between 450 and 650 MB as shown in Fig. 5. The CPU is around 50% for all the experiments. The master node in this experiment was hosting the Kafka queue and more significantly the AUTH topic in which the workers were publishing the results, thus, the CPU utilisation is justified by having the master node writing these events in the authorisation queue.

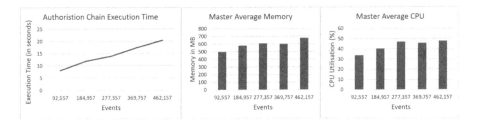

Fig. 5. Execution time, master's memory and CPU utilisation for authorisation.

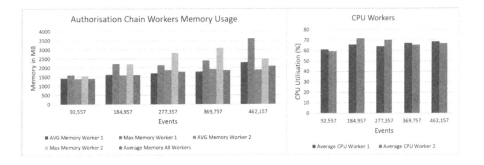

Fig. 6. Workers' memory consumption and CPU utilisation for authorisation.

The average memory consumption and CPU utilisation for the workers is shown in Fig. 6. In this scenario, both workers requiring increasing amount of

---

[4] AMD Opteron(tm) Processor 4226 – 6-cores @ 2.7 GHz, 4 × 16 GB DD3 1066 MHz RAM.

memory for each scenario from the operating system while the CPU utilisation is between 60–70%. The CPU utilisation and memory consumption is similar across the cluster's workers which shows even distribution of work.

4.2   Analytics Scalability Evaluation

We implemented a scenario in which a list of the top products that users browsed within a specific time window is produced. The simulator was instantiated with a varying number of users each of which was randomly navigating a number of products. Navigation of the catalogue has a result of generating one TyphonQL "select" query (e.g., *from Product p select p where* $p.id$ = '...') each time a product page was visited.

The implemented analytics scenario, consumes only those events (i.e., select events on the table Product). As it might be the case that users in real deployments might exploit such an analytics scenario to promote their products (i.e., by visiting their product page repeatedly), we were amending the *slots* attribute of the PreEvent object linked to the PostEvent object that our simulator generated with the id of the user that requested the execution of the command. Such information can be taken for example from the query where the session user id is passed as a comment to the produced query. We produced an increasing number of events which were given as input into the analytics architecture. The analytics code was deployed in the same cluster configuration as described in the evaluation of the authorisation chain.

The time needed is shown in Fig. 7. The graph shows again linear scalability and our architecture does not add any bottleneck. The CPU utilisation and memory consumption for the master node (see Fig. 7) remain quite low as in this experiment the workers are only reading from the POST queue hosted in the master and thus the master is not required to perform any writes to the authorisation queue.

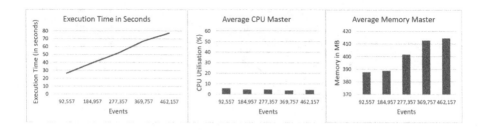

Fig. 7. Execution time, master's memory and CPU for the analytics experiment.

The workers' average and max memory consumption and the average CPU utilisation for the five simulated scenarios are shown in Fig. 8. The workers are using above 80% of the available processing power on average across the five different scalability scenarios. Also, the JVM is claiming all the necessary memory (especially in the last 4 of the five scenarios) but is not running out of memory which is explained by the Java garbage collector replacing unnecessary memory when needed. The load balance is equally split among the workers both in terms of CPU utilisation and memory usage which demonstrates that the workload is shared equally in the cluster.

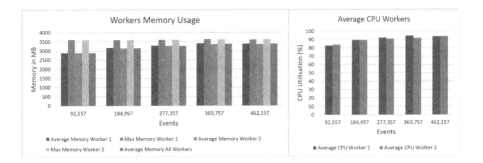

Fig. 8. Workers' memory consumption in the analytics scalability experiment.

5   Related Work

Different systems [8] have been proposed to capture database related evens to mostly allow replication or migration of databases. Connectors (e.g., KafkaConnect [2]) are registered to databases' specific mechanisms to extract already stored data and identify changes. Approaches like Maxwell's Deamon [12] and Oracle GoldenGate [7] monitor the database's log (i.e., binlog) to extract events but these are restricted to use only on relational databases. Debezium [3] offers an event capturing mechanism that supports both relational and non-relational databases. However, it only captures changing commands (i.e., insert, update, delete) and not select queries while it supports a limited number of databases. The Confluent platform [1] is a real-time event streaming application. It supports over 100 connectors to databases and filesystems each of which support different level of granularity of the events that can be captured.

The aforementioned approaches are either limited by the support for a specific set of database types or the amount of processible information. In addition, some of them require duplication of data or storing of unrelated events (e.g., the SQL binlog stores, except the DML commands, DDL commands, too). All of the approaches require the use (and development in case it is not available) of a custom connector for every database and database type in the system. Except for the fact that such connectors might not be possible to be implemented if the database does not offer a related mechanism, the different connectors can acquire different levels of information based on what information the database can offer. In addition, these connectors act separately in each database. If a single TyphonQL command affects more than one database, a common scenario in polystores, then the matching of these events is a difficult - if possible at all - task. Finally, to the best of our knowledge, none of the approaches offer a pre-execution event capturing and authorisation mechanism.

The latter can be achieved with the use of database triggers. However, not all databases allow the execution of custom logic *before* the execution of the commands thus such a feature can be used with some of the databases in the polystore. Most allow the use of triggers after the actual execution of the command. However, this comes with the drawback of having to define specific triggers for each type of command and table/document affected separately which does not allow the creation of a single event that contains all the information needed for the extraction of analytics if a single polystore command affects multiple tables/documents within the same database and across the different databases.

Our approach offers both a before and after execution event capturing mechanism. Authorisation and analytics tasks have access to the data and databases the command affects, no matter if the latter had impact on multiple entities and different types of databases as it is based on a unified syntax (that of TyphonQL). Also, the latter allows future support of event capturing for any new database added to the polystore without requiring developers to implement specific database event capturing/triggering mechanism. Finally, in the case of migration of data from one type of database to another, the authorisation/analytics tasks do not need to be redeveloped to use the database-specific event triggering syntax.

6   Conclusions and Future Work

In this paper we presented a distributed architecture for analytics based on polystore queries. We also presented a pre-execution authorisation mechanism. Finally, the scalability of both the authorisation and the analytics components of the proposed architecture is evaluated. In future work, we will explore if authorisation tasks can be re-arranged automatically in the chain. Tasks that reject a higher proportion of events or require less time to execute could be positioned earlier in the chain. Machine learning can be used to identify the most efficient chains based on different features among those described (i.e., execution time and rejection rate). Finally, applying capture data change (CDC) mechanisms to further facilitate the extraction of analytics would be of interest.

Acknowledgements. This work is funded by the European Union Horizon 2020 TYPHON project (#780251).

References

1. Confluent Inc.: Confluent: Apache Kafka and Event Streaming Platform for Enterprise. https://www.confluent.io/
2. Confluent.io: Kafka Connect. https://docs.confluent.io/current/connect/index.html

3. Debezium Community: Debezium. https://debezium.io/
4. Garg, N.: Apache Kafka. Packt Publishing Ltd., Birmingham (2013)
5. Hueske, F., Kalavri, V.: Stream Processing with Apache Flink: Fundamentals, Implementation, and Operation of Streaming Applications. O'Reilly Media, Newton (2019)
6. Kolovos, D., et al.: Domain-specific languages for the design, deployment and manipulation of heterogeneous databases. In: 2019 IEEE/ACM 11th International Workshop on Modelling in Software Engineering (MiSE), pp. 89–92. IEEE (2019)
7. Oracle Corporation: Real-time access to realtime Information, Oracle White Paper (2015)
8. Rooney, S., et al.: Kafka: the database inverted, but not garbled or compromised. In: 2019 IEEE International Conference on Big Data (Big Data), pp. 3874–3880. IEEE (2019)
9. Strimzi: Strimzi - Apache Kafka on Kubernetes. https://strimzi.io/
10. The Apache Software Foundation: Apache Flink Clusters and Deployment. https://ci.apache.org/projects/flink/flink-docs-release-1.11/ops/deployment/
11. The Apache Software Foundation: Apache Flink Side Outputs. https://ci.apache.org/projects/flink/flink-docs-stable/dev/stream/side_output.html
12. ZenDesk: Maxwell's Daemon. https://maxwells-daemon.io/

Towards Data Discovery by Example

El Kindi Rezig[1]([✉]), Allan Vanterpool[1,2], Vijay Gadepally[3], Benjamin Price[3], Michael Cafarella[1], and Michael Stonebraker[1]

[1] MIT, Cambridge, USA
{elkindi,michjc,stonebraker}@csail.mit.edu
[2] United States Air Force, Washington, D.C., USA
allan.vanterpool@us.af.mil
[3] MIT Lincoln Laboratory, Lexington, USA
{vijayg,ben.price}@ll.mit.edu

Abstract. Data scientists today have to query an avalanche of multi-source data (e.g., data lakes, company databases) for diverse analytical tasks. Data discovery is labor-intensive as users have to find the right tables, and the combination thereof to answer their queries. Data discovery systems automatically find and link (e.g., joins) tables across various sources to aid users in finding the data they need. In this paper, we outline our ongoing efforts to build a data discovery by example system, *DICE*, that iteratively searches for new tables guided by user-provided data examples. Additionally, *DICE* asks users to validate results to improve the discovery process over multiple iterations.

Keywords: Data preparation · Data discovery · Data integration · Data cleaning

1 Introduction

Developing an end-to-end analytic pipeline consists of a number of different operations on the data. One component that is widely recognized as a bottleneck for data scientists, is the time spent discovering and preparing the data needed for their analyses [8,9]. A key step in prepartion is finding and linking the relevant datasets one would need for the task at hand, which when performed manually tends to be both onerous and error-prone [2].

Due to the importance of data discovery, there have been several recent efforts to facilitate it [2,3,6]. Aurum[2] exposes a query API for users to query a graph representation of the data, which captures column similarity and PK-FK relationships. However, the burden is still on the user to write the code to discover and collect the relevant data. In another approach, DoD [4] asks users to provide a schema of the view they are looking for, and the system attempts to find the required joins necessary to produce it. Auto-Join [11] performs transformations on data columns to make them match other key columns and hence make them joinable. However, Auto-Join does not deal with querying the produced data.

© Springer Nature Switzerland AG 2021
V. Gadepally et al. (Eds.): Poly 2020/DMAH 2020, LNCS 12633, pp. 66–71, 2021.
https://doi.org/10.1007/978-3-030-71055-2_6

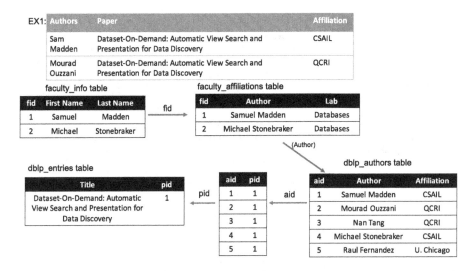

Fig. 1. Example PK-FK join path across different tables

Consider the tables in Fig. 1. Suppose we have the following query Q_1: "Get me the list of papers co-authored by DB faculty at CSAIL and non-CSAIL authors". If the user knows every table in the data lake, then, it's straightforward to write a SQL query to answer the query. However, the number of tables in data lakes can be unbounded, which renders manual inspection impractical. A data discovery system that can aid in finding the PK-FK join paths (arrows in Fig. 1) makes it easier for a user to know which tables join to what others, and then the user can write queries on top of that pre-processed data. However, the user still needs to write the queries, which in many cases can be time-consuming and cumbersome (i.e., the user does not know all the tables, and the relationships thereof), based on the level of depth and complexity of those relationships.

We are working on $DICE$ (Data Discovery by Example), a system that enables data discovery by example, wherein the user provides a set of desired records as examples, and the system then automatically fetches the relevant tables/columns, which would require doing joins between columns of different tables. For instance, to express Q_1, the user could provide two example records of the desired output (EX1 in Fig. 1) from which $DICE$ extracts values and properties (e.g., $DICE$ would notice that the values in affiliation can be different in EX1) to look up relevant tables in the data lake, and then construct join paths. The lookup function in DICE supports both exact and similarity matching.

From our ongoing engagement with one large organization, we have learned that being able to search data by example would be a highly desirable improvement for data analysts, as opposed to their current method of writing queries which requires explicit knowledge of the underlying data organization.

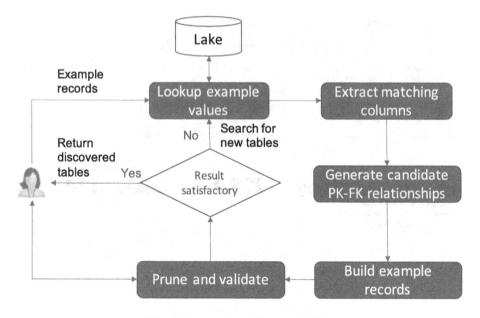

Fig. 2. General workflow of DICE

2   System Overview

Figure 2 outlines the overall workflow of *DICE*. In a nutshell, *DICE* does not perform data discovery in one shot, but instead is an interactive system. DICE interacts with the user in two ways: (1) by requesting example records; and (2) by asking them to validate constructed join paths. After each iteration the user can decide to either stop the search, or prompt *DICE* to look for more candidate tables and join paths.

As mentioned above, the first step is for the user to provide a table (see EX1 in Fig. 1) as input containing a set of example records. After extracting the values and properties from the example, *DICE* fetches relevant tables from the data lake. *DICE* then attempts to construct PK-FK relationships from the columns of the extracted tables. *DICE* then constructs sample records from the selected join paths, presents them to the user and based on the user's (Yes/No) feedback on each record, fetches more tables or stops the search if the user is satisfied with the results. The user may also supply additional examples after each iteration. Given this design we note the following challenges:

- **Extracting the values/properties from the user's example records:** Interpreting example records can be ambiguous. For instance, in Fig. 1, the example EX1 could mean: "get me all papers co-authored by DB CSAIL faculty and non-CSAIL authors" or "get me all papers written by anybody in Databases" or "get me all papers co-authored by Samuel Madden and Mourad Ouzzani". Therefore, primitives are needed so the user can express their "intent" over the provided examples.

- **Initial search region:** Initially, *DICE* has to fetch a set of tables, and try to link them together. If too many tables match the user's examples, choosing the *correct* subset of tables for the user to consider first, can be challenging. The initial search region (first tables to consider in the lake) is key because *DICE* builds on it for its subsequent searches by expanding it to include more tables.
- **Minimizing the user's input:** *DICE* needs to actively involve the user by selecting "interesting" and relevant records for validation. Additionally, *DICE* will need to be able to ask the user for more examples if the current examples are too broad or limited for the search. It is crucial to minimize the user's interactions while attempting to maximize the value of the user's feedback, when needed.

2.1 Knowledge Graphs

In addition to the relational data model, we are also looking into how *DICE* could be used in other data models. For instance, a knowledge graph encodes entities (nodes) and the properties that connect them (edges). In this case, *DICE* is not looking for PK-FK relationships, but for any relationship between nodes in the graph (paths). This makes the problem more challenging because *DICE* has to select only relevant paths to consider (which would require a measure of *interestingness* attributed to different paths). Once determined, *DICE* can then build records from those paths to present samples to the user.

3   Benefits of *DICE*

Using a discovery tool such as *DICE*, can provide users with a number of benefits:

- **Simplifying data discovery:** By allowing users to specify examples of the data output they would like, *DICE* greatly reduces the number of iterations between the user and system. The user helps define the high-level data discovery task and the system provides viable alternative outputs for the user to choose between, allowing for a quick narrow-down to the most relevant alternative.
- **Improving data aggregate sensitivity:** In many computing environments, while individual records may not present concerns, aggregation of different types of data can lead to policy or security violations. For example, integration of different non-sensitive datasets may lead to sensitive outputs. Consider a medical scenario in which a user is trying to integrate data from *Table 1* and *Table 2*. Suppose *Table 1* consists of columns (PatientID, Patient Name, Patient Gender) and *Table 2* consists of columns (PatientID, Patient Date of Birth, Patient Age). If a researcher is interested in looking at the distribution of gender and age within the datasets, they may attempt to do a "join" on Table 1 and Table 2 using the PatientID. Very often, combining Patient Name

and Date of Birth can lead to sensitive Personally Identifiable Information (PII). Using a tool like *DICE*, the user could specify *a priori* that they are looking only for (Gender, Age) and *DICE* can filter data appropriately before aggregating.

- **Improved policy compliance:** In many environments, data is distributed across different systems with different owners. These data owners may not wish to provide unfettered access to traditional data discovery tools. In which case, they could expose only a subset of their data or metadata about their system. *DICE* could use this limited information to construct "safely" pruned datasets for the data scientist. If any of these are of interest, the data scientist can then reach out to data owners for greater access.

- **Working well within polystore or multistore environments:** Data distributed across heterogeneous systems, as seen in polystores [5] and multistore environments [10], pose new challenges for data discovery. A tool such as *DICE* would improve data integration across these heterogeneous systems. For example, the applications described in [1,7], leverage data stored in disparate data management systems. For a data scientist to find data of interest, they must manually query each individual system (or write a polystore query that speaks to multiple, different backend systems). In this case as well, *DICE* could create these queries for the data scientist in order to help them find the specific data products of interest.

4   Conclusion

DICE is an interactive, user-in-the-loop, data discovery by example system. In this paper we presented a quick walkthrough of the workflow of *DICE* as well as the challenges associated with its implementation. We are currently working with one large-scale organization to implement *DICE* on their data lake, but believe that *DICE* may have widespread value and application to other organizations as data storage continues to grow in both complexity and distribution.

Acknowledgement. Research was sponsored by the United States Air Force Research Laboratory and was accomplished under Cooperative Agreement Number FA8750-19-2-1000. The views and conclusions contained in this document are those of the authors and should not be interpreted as representing the official policies, either expressed or implied, of the United States Air Force or the U.S. Government. The U.S. Government is authorized to reproduce and distribute reprints for Government purposes notwithstanding any copyright notation herein.

References

1. Elmore, A.J., et al.: A demonstration of the BigDAWG polystore system. Proc. VLDB Endow. **8**(12), 1908 (2015)
2. Fernandez, R.C., Abedjan, Z., Koko, F., Yuan, G., Madden, S., Stonebraker, M.: Aurum: a data discovery system. In: 34th IEEE International Conference on Data Engineering, ICDE 2018, Paris, France, 16–19 April 2018, pp. 1001–1012. IEEE Computer Society (2018). https://doi.org/10.1109/ICDE.2018.00094

3. Fernandez, R.C., et al.: Seeping semantics: linking datasets using word embeddings for data discovery. In: 34th IEEE International Conference on Data Engineering, ICDE 2018, Paris, France, 16–19 April 2018, pp. 989–1000. IEEE Computer Society (2018). https://doi.org/10.1109/ICDE.2018.00093

4. Fernandez, R.C., Tang, N., Ouzzani, M., Stonebraker, M., Madden, S.: Dataset-on-demand: automatic view search and presentation for data discovery. CoRR abs/1911.11876 (2019). http://arxiv.org/abs/1911.11876

5. Gadepally, V., et al.: The BigDAWG polystore system and architecture. In: 2016 IEEE High Performance Extreme Computing Conference (HPEC), pp. 1–6. IEEE (2016)

6. Halevy, A.Y., et al.: Goods: organizing Google's datasets. In: Özcan, F., Koutrika, G., Madden, S. (eds.) Proceedings of the 2016 International Conference on Management of Data, SIGMOD Conference 2016, San Francisco, CA, USA, 26 June–01 July 2016, pp. 795–806. ACM (2016). https://doi.org/10.1145/2882903.2903730

7. Mattson, T., Gadepally, V., She, Z., Dziedzic, A., Parkhurst, J.: Demonstrating the BigDAWG polystore system for ocean metagenomics analysis. In: CIDR (2017)

8. Rezig, E.K., et al.: Data civilizer 2.0: a holistic framework for data preparation and analytics. PVLDB **12**(12), 1954–1957 (2019). https://doi.org/10.14778/3352063.3352108. http://www.vldb.org/pvldb/vol12/p1954-rezig.pdf

9. Rezig, E., Cafarella, M., Gadepally, V.: Technical report: an overview of data integration and preparation (2020)

10. Tan, R., Chirkova, R., Gadepally, V., Mattson, T.G.: Enabling query processing across heterogeneous data models: a survey. In: 2017 IEEE International Conference on Big Data (Big Data), pp. 3211–3220. IEEE (2017)

11. Zhu, E., He, Y., Chaudhuri, S.: Auto-join: joining tables by leveraging transformations. Proc. VLDB Endow. **10**(10), 1034–1045 (2017). https://doi.org/10.14778/3115404.3115409. http://www.vldb.org/pvldb/vol10/p1034-he.pdf

The Transformers for Polystores - The Next Frontier for Polystore Research

Edmon Begoli$^{(\boxtimes)}$, Sudarshan Srinivasan, and Maria Mahbub

Oak Ridge National Laboratory, Oak Rdige, TN 37831, USA
{begolie,srinivasans,mahbubm}@ornl.gov

Abstract. What if we could solve one of the most complex challenges of polystore research by applying a technique originating in a completely different domain, and originally developed to solve a completely different set of problems? What if we could replace many of the components that make today's polystore with components that only understand query languages and data in terms of matrices and vectors? This is the vision that we propose as the next frontier for polystore research, and as the opportunity to explore attention-based transformer deep learning architecture as the means for automated source-target query and data translation, with no or minimal hand-coding required, and only through training and transfer learning.

Keywords: Polystore · Transformers attention-based neural networks · Deep learning

1 Introduction

In 2005, Stonebraker and Çetintemel [17] posited that the time of "One Size Fits All" database management systems is over. The era of "Big Data" brought the challenge of a variety of formats, large volumes, and specialized systems (relational, document, graph, etc.) required to manage different data domains.[1] As a result of this need for diversification, we have seen a rise of new data management systems and styles, successful new special-purpose technologies, and the research efforts, such as polystores, that bring back the idea of federated style databases [8]. Polystore research has inspired numerous new directions and solutions, which we survey in the next section.

This idea, however, faces similar issues that the original federated database idea faced. There is a challenge of creating a uniform interface against all sources

---

[1] The United States Government retains and the publisher, by accepting the article for publication, acknowledges that the United States Government retains a non-exclusive, paid-up, irrevocable, world-wide license to publish or reproduce the published form of this manuscript, or allow others to do so, for United States Government purposes. The Department of Energy will provide public access to these results of federally sponsored research in accordance with the DOE Public Access Plan (http://energy.gov/downloads/doe-public-access-plan).

© Springer Nature Switzerland AG 2021
V. Gadepally et al. (Eds.): Poly 2020/DMAH 2020, LNCS 12633, pp. 72–77, 2021.
https://doi.org/10.1007/978-3-030-71055-2_7

(islands) covered under the polystores while simultaneously maintaining the independence of individual sources' access and manageability. The engineering effort involved in developing and maintaining a unifying layer is significant. This engineering effort usually requires a significant amount of programming. Then, there is a need to perform a source-to-source translation between database query languages (e.g., from SQL to SQL-like dialects), and there is sometimes a need to do a translation between the data sources to produce useful and meaningful results (e.g., translate long narratives to computable summaries).

However, in the field of artificial intelligence, we observed state-of-the-art techniques that enable human-like performance in machine translations, language generation, and other useful transformations. Our experience with the development and use of polystores architectures and related approached, and our exposure to these AI techniques inspire us to propose a new research approach that combines the AI research with polystores research, with a promise that some of these AI techniques could help simplify some of the engineering challenges related to the implementation, use, and maintenance of polystores.

2 Challenges with Current Approaches

As we have discussed already, a canonical polystore architecture uses *shims* for language translation between the native, *island* datastore and a polystores. While this is a convenient and necessary feature to create a user-friendly experience, it is also a complex one, and it comes with some significant challenges. For example, polystore system requires multiple shims to carryout translation from one language to another which requires all the supporting engineering mechanisms to translate one database language to another. While this is a sound, albeit labor-intensive approach, the ideal solution would be to automate the translation between the two languages (or two dialects of the same database language), and hence remove the need for significant software engineering effort. In the next sections, we discuss what this solution could be.

3 Natural Language Processing with Transformers

In recent years, deep learning has revolutionized the field of natural language processing (NLP). While many deep learning architectures have been used for processing natural language (such as convolutional neural networks (CNN) [9] long short-term memory (LSTM) networks [12], Temporal Convolutional Networks (TCN) [11]), attention based networks [7] have been at the forefront of deep learning based NLP models. The attention mechanism is a part of a neural architecture that enables to dynamically highlight relevant features of a sequence of textual elements. The transformer architecture [18] effectively uses attention for long sequences dispensing recurrence and convolutions entirely. Transformers have been been very successful in various NLP applications [21].

The transformer architecture along with the notion of transfer learning via pre-trained language models was effectively used to create the Bidirectional

Encoder Representations from Transformers (BERT) [4]. At the time of its publication, BERT broke several state-of-art results on quintessential NLP tasks on widely used datasets. Since then, there have many derivatives of BERT, many of them catering to specific scenarios. Examples include GPT [15,16], GPT-3 [1], Transformer-XL [3], XLNet [22], RoBERTa [13], and many others. For more information please see [20].

GPT-3 [1] is the most recently released model (July 2020). This is an autoregressive language model that was trained with large number of parameters (175 billion). Its abilities include translation, question-answering, Cloze tasks [14], and several tasks that require on-the-fly reasoning or domain adaption. Furthermore, due to large amount of training and parameters, this model is capable of few-shot and one-shot training [19] on certain tasks. Few-shot training refers to fine-tuning the language model on a few task specific example, while one-shot refers to fine-tuning with one example. Interestingly, it was shown that transformer architectures (e.g., Transformer) can be used to generate source code for programming languages [10]. We believe this aspect of pre-trained and fine-tuned transformers in general could be the subject of the next stage of Polystore research.

4   The Role for Transformers in Polystore Research

We propose to use this attention-based transformer architecture as a means of augmenting shim functionality. We believe that a neural machine translation system that uses transformer architecture can be trained to *translate* polystore language queries into the underlying island language query. In particular, we believe that the advanced models such as GPT-3 transformer with very minimal fine-tuning could be used to either augment the translation made by or completely replace the shim component of the polystore framework (Fig. 1).

For example, BigDAWG's *shim* can translate the *SELECT * FROM table* SQL query to *SCAN(table)* in SciDB, or it can translate one dialect of SQL to another. In fact, we posit that translation of one declarative language such as SQL to another, is a lesser challenge than translating English to Finnish, or French to Mandarin, where the grammatical differences, and possible variations are far greater than between the declarative database languages. Yet, we have seen transformers achieve the near-human, state-of-the-art results [2].

We further hypothesize that GPT-3-like transformer could be employed for polystore query translation with minimal or without any fine-tuning, and we base this hypothesis on above mentioned results in related work [2,10].

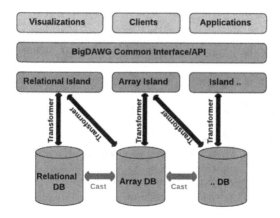

Fig. 1. Our Concept of BigDAWG [5,6] with Transformers

5 Future Work

An incorporation of transformer architectures into polystore research is an exciting and challenging idea. Given its novelty it is hard to exactly pin down all the possible directions that this research can take. For that reason, we discuss here areas that we observe as the ones that we plan to undertake, as well as the ones that are the most obvious – at least to us.

We expect that transformer architectures will play a role in two areas of polystore research, namely:

- use of transformers as source-to-source translators, and
- use of transformers as data translators.

While the current work has already shown that transformers can do a source-to-source translation between programming languages (transpilation), we expect that other transformer functions, such as auto-summarization, and other forms of transformations, could play a role in polystore research. For example, we could see in the future transformers used to summarize text into sentences, which could be served as columnar results, transform semi-structured data into structured tabular form, translate data in one native language to another, and many others.

There are perhaps too many ideas and future directions, and we see that as a good state. We hope that this paper will serve as an inspiration for durable and broad research into how one breakthrough technology can benefit the other.

Acknowledgments. This work has been in part co-authored by UT- Battelle, LLC under Contract No. DE-AC05-00OR22725 with the U.S. Department of Energy. The content is solely the responsibility of the authors and does not necessarily represent the official views of the UT-Battelle, or the Department of Energy.

References

1. Brown, T.B., et al.: Language models are few-shot learners. arXiv e-prints arXiv:2005.14165, May 2020
2. Brown, T.B., et al.: Language models are few-shot learners. arXiv preprint arXiv:2005.14165 (2020)
3. Dai, Z., Yang, Z., Yang, Y., Carbonell, J., Le, Q.V., Salakhutdinov, R.: Transformer-XL: attentive language models beyond a fixed-length context. arXiv e-prints arXiv:1901.02860, January 2019
4. Devlin, J., Chang, M.W., Lee, K., Toutanova, K.: BERT: pre-training of deep bidirectional transformers for language understanding. arXiv e-prints arXiv:1810.04805, October 2018
5. Duggan, J., et al.: The BigDAWG polystore system. ACM SIGMOD Rec. **44**(2), 11–16 (2015)
6. Gadepally, V., et al.: The BigDAWG polystore system and architecture. In: 2016 IEEE High Performance Extreme Computing Conference (HPEC), pp. 1–6. IEEE (2016)
7. Galassi, A., Lippi, M., Torroni, P.: Attention in natural language processing. arXiv e-prints arXiv:1902.02181, February 2019
8. Hsiao, D.K.: Federated databases and systems: part I–a tutorial on their data sharing. VLDB J. **1**(1), 127–179 (1992)
9. Kim, Y.: Convolutional neural networks for sentence classification. In: Proceedings of the 2014 Conference on Empirical Methods in Natural Language Processing (EMNLP), pp. 1746–1751 (2014)
10. Lachaux, M.A., Roziere, B., Chanussot, L., Lample, G.: Unsupervised translation of programming languages. arXiv preprint arXiv:2006.03511 (2020)
11. Lea, C., Flynn, M.D., Vidal, R., Reiter, A., Hager, G.D.: Temporal convolutional networks for action segmentation and detection. In: Proceedings of the IEEE Conference on Computer Vision and Pattern Recognition, pp. 156–165 (2017)
12. Liu, G., Guo, J.: Bidirectional LSTM with attention mechanism and convolutional layer for text classification. Neurocomputing **337**, 325–338 (2019)
13. Liu, Y., et al.: RoBERTa: a robustly optimized BERT pretraining approach. arXiv e-prints arXiv:1907.11692, July 2019
14. Neville, M.H., Pugh, A.: Context in reading and listening: variations in approach to cloze tasks. Read. Res. Q. 13–31 (1976)
15. Radford, A., Narasimhan, K., Salimans, T., Sutskever, I.: Improving language understanding by generative pre-training (2018)
16. Radford, A., Wu, J., Child, R., Luan, D., Amodei, D., Sutskever, I.: Language models are unsupervised multitask learners. OpenAI Blog **1**(8), 9 (2019)
17. Stonebraker, M., Cetintemel, U.: "One size fits all": an idea whose time has come and gone. In: 21st International Conference on Data Engineering (ICDE 2005), pp. 2–11. IEEE (2005)
18. Vaswani, A., et al.: Attention Is All You Need. arXiv e-prints arXiv:1706.03762, June 2017
19. Vinyals, O., Blundell, C., Lillicrap, T., Wierstra, D., et al.: Matching networks for one shot learning. In: Advances in Neural Information Processing Systems, pp. 3630–3638 (2016)
20. Wolf, T., et al.: HuggingFace's transformers: state-of-the-art natural language processing. arXiv e-prints arXiv:1910.03771, October 2019

21. Wolf, T., et al.: Transformers: state-of-the-art natural language processing. arXiv preprint arXiv:1910.03771 (2019)
22. Yang, Z., Dai, Z., Yang, Y., Carbonell, J., Salakhutdinov, R.R., Le, Q.V.: XLNet: generalized autoregressive pretraining for language understanding. In: Advances in Neural Information Processing Systems, pp. 5753–5763 (2019)

DMAH 2020: COVID-19 Data Analytics and Visualization

Open-World COVID-19 Data Visualization [Extended Abstract]

Hyunseung Hwang and Steven Euijong Whang$^{(\boxtimes)}$

Korea Advanced Institute of Science and Technology, Daejeon, South Korea
{aguno,swhang}@kaist.ac.kr

Abstract. As COVID-19 becomes a dangerous pandemic worldwide, there is an urgent need to understand all aspects of it through data visualization. As part of a larger COVID-19 response by KAIST, we have worked with students on generating interesting COVID-19 visualizations including demographic trends, patient behaviors, and effects of mitigation policies. A major challenge we experienced is that, in an *open world setting* where it is not even clear which datasets are available and useful, generating the right visualizations becomes an extremely tedious process. Traditional data visualization recommendation systems usually assume that the datasets are given, and that the visualizations have a clear objective. We contend that such assumptions do not hold in a COVID-19 setting where one needs to iteratively adjust two moving targets: deciding which datasets to use, and generating useful visualizations with the selected datasets. We thus propose interesting research challenges that can help automate this process.

1 Introduction

The COVID-19 pandemic is widely considered as one of the world's biggest challenges since World War II. Even as some countries are successful in overcoming the first wave of this pandemic, they are bracing for a second wave that is likely to come towards the end of the year. Hence, there is an urgent need to understand all aspects of COVID-19. We not only need to develop vaccines, but also understand how the disease spreads, how people are reacting, how to mitigate COVID-19, and more. Data visualization is commonly used to provide decision support for policy making.

Public health 2.0 has never been more important where we can utilize a growing list of public and restricted, but accessible data sources. Some well-known sources include the WHO website, Johns Hopkins Coronavirus Resource Center, the Centers for Disease Control and Prevention (CDC), and Kaggle. South Korea is one of the leading countries for combating COVID-19, and there is a Korean version of the CDC (KCDC) that shows various statistics about COVID-19 patients. In addition, the Korean government supports access to various datasets through data safe zones (e.g., Korea Telecom mobile data) and

Supported by a Google AI Focused Research Award.

V. Gadepally et al. (Eds.): Poly 2020/DMAH 2020, LNCS 12633, pp. 81–84, 2021.
https://doi.org/10.1007/978-3-030-71055-2_8

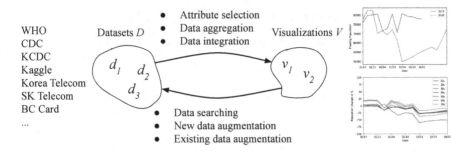

Fig. 1. In open-world data visualization, finding the right datasets and visualizations is repeated until the visualizations are considered insightful to the analyst.

funds data voucher projects where companies process and sell datasets that help COVID-19 analysis (e.g., BC Card credit card data). There are also various data visualization competitions for COVID-19.

Ironically, all these data sources make data visualization extremely challenging, as there is a large search space of combinations of datasets and possible visualizations on them. Traditional data visualization systems are not sufficient because they usually assume the input datasets are fixed, and the visualizations have specific performance goals. In reality, we are in an *open-world setting* where datasets are continuously evolving (e.g., KCDC may update its dataset or introduce a new one) and the helpful visualizations may change as new events occur (e.g., a new policy for social distancing). Hence, open-world data visualization involves multiple iterations of two moving targets: searching and integrating potentially useful datasets [1] and generating the right visualizations on them [3]. Both topics have been studied individually under various names, but have seldom been studied together within the same workflow.

We suggest research challenges that can help automate open-world data visualization as illustrated in Fig. 1. Given datasets, generating visualizations involves selecting attributes, aggregating data, and integrating data. Once visualizations have been generated, the data analyst may improve them by searching for new datasets and augmenting the existing ones. We also provide case studies of working with independent study students to generate useful COVID-19 visualizations. While we were able to obtain a number of interesting visualizations, the iterative trial and error with students literally took weeks. In hindsight, this process was tedious and not fast enough to generate visualizations to be useful during the first wave of the pandemic. For example, Korea initially experienced a shortage of face masks, but by the time we figured out how to estimate the supply and demand of masks per region, the problem was largely solved. However, our efforts are expected to be useful for the possible second wave.

2   Research Challenges

The goal of open-world data visualization is to choose a set of datasets $D = \{d_1, d_2, \ldots d_n\}$ and a set of visualizations $V = \{v_1, v_2, \ldots v_m\}$ that provide the best insights of COVID-19. Whether a visualization is insightful can be determined manually by the data analyst. If the analyst is not satisfied, she can generate new visualizations on the current datasets or search for more datasets that complement the current ones. Even if the visualizations look good now, there may be new events that prompt the analyst to generate better ones later.

We identify research challenges that occur when exploring visualizations and searching datasets. Some of the challenges are not new in data analytics, but we would like to tailor them to an open-world visualization setting.

1. Exploring visualization candidates given a set of datasets $(D \to V)$
   (a) *Attribute selection*: Finding the right attributes to show in visualizations is a time-consuming process where the analyst has to identify which combinations of attributes produce interesting visualizations.
   (b) *Data aggregation*: Determining the granularity (e.g., daily/weekly/monthly) of analyses and deciding which attributes (e.g., age, region) to group on. This process can be manual and repetitive.
   (c) *Data integration*: Aligning possibly-inconsistent attributes so the datasets can be joined. The integration may be repeated as the analyst finds new datasets to augment existing visualizations.
2. Searching new datasets to improve the visualizations $(V \to D)$
   (a) *Data searching*: The analyst may realize the given datasets are unsuitable for generating the desired visualization and may need to search new datasets from scratch. While there are existing dataset search tools, they do not necessarily cover recently-added COVID-19 data sources.
   (b) *New data augmentation*: The analyst may want to add new visualizations on top of existing ones and search for new datasets that contain the needed information. Unlike searching from scratch, the new datasets may need to be integrated with the existing ones.
   (c) *Existing data augmentation*: This case is similar to new data augmentation, but the emphasis is more on supplementing any missing values of existing datasets and visualizations.

3   Case Studies

We highlight the data visualization experiences of two analysts[1] among others. Both analysts ran into most of the research challenges above, and we specify when the new or existing data augmentations occur below.

Analyst A wanted to visualize types of locations visited by COVID-19 patients over time. Initially, Analyst A used a patient route dataset provided by KCDC. This dataset contains for each patient a list of locations where each location has a latitude, longitude, and location type. After the initial visualization, however, Analyst A noticed that many location types were not categorized and had "etc." values as shown in Fig. 2a. Analyst A then decided to utilize a public building information database to fill in the missing categories (i.e., existing data

---

[1] Credits go to Beomsik Park and Jaeyoung Park.

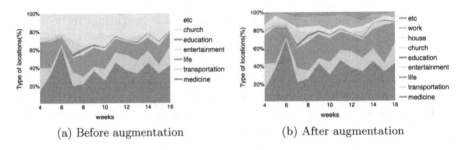

(a) Before augmentation   (b) After augmentation

Fig. 2. (a) Types of locations visited by COVID-19 patients over time. (b) After augmenting the "etc." values (yellow region) using additional building information. (Color figure online)

augmentation) as shown in Fig. 2b. This operation required a non-trivial conversion from latitude/longitude values to building addresses using Naver Maps before joining with the building information database.

Analyst B wanted to see how significant COVID-19 events influence people's behaviors. Initially, Analyst B focused on subway usage data provided from the Seoul Open Data Plaza website. The subway population significantly decreases after two events: right after the first COVID-19 patients were diagnosed in Korea and after a major outbreak by members of Shincheonji. Analyst B then wanted to compare this visualization with the more general floating population data by obtaining a new dataset provided by SK Telecom (i.e., new data augmentation). As a result, while the floating population also decreases after the first COVID-19 patient diagnosis, it does not decrease much after the outbreak. We suspect that people learned how to cope with COVID-19 better and, instead of taking the subway, drove themselves during the outbreak.

4   Discussion

COVID-19 is here to stay, and understanding it through data visualizations will only become more important. In addition to the existing approaches for closed-world data visualization, we hope the research community will tackle the novel problem of accelerating open-world data visualization. Recently, deep learning approaches [2] have been used to automate data visualizations by determining which plot types are suitable for visualizing which attributes. An interesting direction is to expand this approach to jointly perform dataset searching as well.

References

1. Brickley, D., Burgess, M., Noy, N.F.: Google dataset search: building a search engine for datasets in an open web ecosystem. In: WWW, pp. 1365–1375. ACM (2019)
2. Hu, K.Z., Bakker, M.A., Li, S., Kraska, T., Hidalgo, C.A.: Vizml: a machine learning approach to visualization recommendation. In: CHI, p. 128 (2019)
3. Wongsuphasawat, K., et al.: Voyager 2: augmenting visual analysis with partial view specifications. In: CHI, pp. 2648–2659. ACM (2017)

DMAH 2020: Deep Learning based Biomedical Data Analytics

Privacy-Preserving Knowledge Transfer with Bootstrap Aggregation of Teacher Ensembles

Hong-Jun Yoon[1(✉)] , Hilda B. Klasky[1] , Eric B. Durbin[2], Xiao-Cheng Wu[3],
Antoinette Stroup[4], Jennifer Doherty[5], Linda Coyle[6], Lynne Penberthy[7],
Christopher Stanley[1], J. Blair Christian[1], and Georgia D. Tourassi[8]

[1] Computational Sciences and Engineering Division, Oak Ridge National Laboratory,
Oak Ridge, TN 37830, USA
{yoonh,klaskyhb,stanleycb,christianjb}@ornl.gov
[2] College of Medicine, University of Kentucky, Lexington, KY 40536, USA
ericd@kcr.uky.edu
[3] Louisiana Tumor Registry, School of Public Health, Louisiana State University
Health Sciences Center, New Orleans, LA 70112, USA
XWu@lsuhsc.edu
[4] New Jersey State Cancer Registry, Rutgers Cancer Institute of New Jersey,
New Brunswick, NJ 08901, USA
nan.stroup@rutgers.edu
[5] Utah Cancer Registry, Huntsman Cancer Institute, University of Utah, Salt Lake
City, UT 84132, USA
Jen.Doherty@hci.utah.edu
[6] Information Management Services Inc., Calverton, MD 20705, USA
coylel@imsweb.com
[7] Surveillance Research Program, Division of Cancer Control and Population
Sciences, National Cancer Institute, Bethesda, MD 20814, USA
lynnepenberthy.schumacher-penberthy@nih.gov
[8] National Center for Computational Sciences, Oak Ridge National Laboratory,
Oak Ridge, TN 37830, USA
tourassig@ornl.gov

Abstract. There is a need to transfer knowledge among institutions
and organizations to save effort in annotation and labeling or in enhancing task performance. However, knowledge transfer is difficult because of
restrictions that are in place to ensure data security and privacy. Institutions are not allowed to exchange data or perform any activity that may
expose personal information. With the leverage of a differential privacy

This manuscript has been authored in part by UT-Battelle, LLC, under contract DE-
AC05-00OR22725 with the US Department of Energy (DOE). The US government
retains and the publisher, by accepting the article for publication, acknowledges that
the US government retains a nonexclusive, paid-up, irrevocable, worldwide license to
publish or reproduce the published form of this manuscript, or allow others to do so, for
US government purposes. DOE will provide public access to these results of federally
sponsored research in accordance with the DOE Public Access Plan (http://energy.
gov/downloads/doe-public-access-plan).

© Springer Nature Switzerland AG 2021
V. Gadepally et al. (Eds.): Poly 2020/DMAH 2020, LNCS 12633, pp. 87–99, 2021.
https://doi.org/10.1007/978-3-030-71055-2_9

algorithm in a high-performance computing environment, we propose a new training protocol, Bootstrap Aggregation of Teacher Ensembles (BATE), which is applicable to various types of machine learning models. The BATE algorithm is based on and provides enhancements to the PATE algorithm, maintaining competitive task performance scores on complex datasets with underrepresented class labels.

We conducted a proof-of-the-concept study of the information extraction from cancer pathology report data from four cancer registries and performed comparisons between four scenarios: no collaboration, no privacy-preserving collaboration, the PATE algorithm, and the proposed BATE algorithm. The results showed that the BATE algorithm maintained competitive macro-averaged F1 scores, demonstrating that the suggested algorithm is an effective yet privacy-preserving method for machine learning and deep learning solutions.

Keywords: Data privacy · Privacy-preserving machine learning · Differential privacy · Bootstrap aggregation · Information extraction · Natural language processing

1   Introduction

Data security and privacy are prime topics in the design of artificial intelligence (AI) systems [14,15,17,18]. Domains such as biomedical and health informatics, finance, tax revenue services, and homeland security characteristically use sensitive data that contains personal information about human subjects. For the safety of the data and personal information, exchanging such data among organizations and institutions is strictly controlled to prevent any possible leakage of sensitive human subject information. However, to develop faithful deep learning (DL)-based machine learning (ML) information extraction and classification models, a large amount of data from various data sources is highly desirable. Moreover, some institutions may be limited by having too few training examples to achieve ML/DL models to meet their expectations [10]. Thus, there is a need for ways to transfer knowledge securely among organizations and institutions.

However, current AI and ML-based data processing approaches present security vulnerabilities that can be exploited to leak sensitive details. Exposure of private information can occur as a result of the features captured by DL models. A key feature of DL models is that they equip multiple layers of trainable parameters. They learn by example and extract optimal feature representations to enable higher accuracy. However, the ML/DL training algorithm is domain-agnostic and does not recognize if a feature contains sensitive information.

Privacy-preserving models aim to prevent the (identification and) storage of sensitive information in training data used in ML algorithms. Privacy herein is understood as establishing a differential privacy approach that identifies privacy with a measurable and rigorous mathematical definition [5]. Differential privacy allows companies to collect the data of users without compromising the privacy of the individuals [7]. Differential privacy [6] ensures that the probability distribution of the released statistics is roughly similar without paying attention to the inclusion or non-inclusion of any single member in the study; thus, it provides credible statistics. Applying differential privacy to ML algorithms provides a very strong guarantee that the datasets can be shared across registries without concerns about privacy and confidentiality.

The study described in this document focuses on the application and evaluation of an approach based on the Private Aggregation of Teacher Ensembles (PATE) algorithm [16]. PATE is a modified teacher/student model that includes differential privacy [16]. PATE initiates by working on a set of sensitive data, which is partitioned into different sections that do not overlap. On each partition, any ML model, which in PATE's framework is called a "teacher," is trained. The set of ML models or teachers is called an "assemble." Different and independent learning models can be used in each partition separately. At the inference phase of the algorithm, PATE aggregates the predictions of the teacher assemble. To do so, PATE counts votes, adds Laplace noise to the teachers' answers, and then takes the maximum value. The Laplace noise introduces randomness to protect the privacy of users when the teachers do not have a strong quorum. The final step is to transfer the knowledge from the teacher assemble model to a student model using some public data (unlabeled). The teacher assemble will label some of the unlabeled public data, and the student model will contain a training set that will be used to learn a model and perform predictions. The student model is added to decrease the probability of total privacy loss. In recent years, there have been several refinements to the PATE model; specifically, there have been improvements to the student model part of the algorithm [19].

One limitation that we observed with PATE is that it is too conservative regarding underrepresented and minor classes during the classification process. To address this issue, we propose an enhancement to the PATE framework. We named our approach BATE (Bootstrap Aggregation of Teacher Ensemble). BATE uses bagging (bootstrap aggregating) in high-performance computing (HPC), instead of the data partitioning implemented in the PATE framework. Thus, it yields performance scores for the minor classes at the same time that it ensures differential privacy. We hypothesize that including the bootstrap classification will help improve BATE HPC performance.

In this paper, we performed a feasibility study of the proposed BATE model with the data from four cancer registries. We developed models to extract information on morphological and topographical characteristics of tumors from cancer pathology reports. The cancer pathology dataset was labeled by the cancer/tumor/case codes that met the Surveillance, Epidemiology, and End Results (SEER) case reporting guidelines. We developed multitask convolutional neural network (MT-CNN) models and confirmed that the model was feasible for the cancer pathology report corpus [4]. In this study, we simulated a scenario in which one cancer registry had no gold standard labels and so learned from the other three registries. But there was a restriction that no cancer registry should expose patients' identities and information to others. The results presented in this study support our hypothesis on performance improvement. We show that the performance of the BATE model is superior to that of the PATE model, especially for subsite and histology, those classes suffering from severe class imbalance, and many underrepresented classes.

This paper is organized as follows: Sect. 2 presents related work, and Sect. 3 presents the data and methods used in the study. Section 4 presents the results. Section 5 provides a discussion, limitations, conclusions, and future work.

2   Related Work

Part of the groundwork that established the foundation of PATE was the work on differential privacy on neural networks by Abadi et al. [2]. That study introduced a differential privacy stochastic gradient descent (SGD) algorithm aimed at controlling the influence of the training data stage, specifically in SGD computation. In a subsequent study, PATE was presented by Papernot et al. [16] as an independent approach to a learning algorithm for either teacher or student models, i.e. a black-box approach, and consequently, capable of being applied to other learning methods. PATE improved the accuracy of a private MNIST model from 97% to 98% and the privacy bound from 8 to 1.9 [16]. Note that MNIST is a simple classification task. The following are other variations of the PATE approach:

- A PATE variation called PATE-G was introduced by Abadi et al. [3]. PATE-G implements generative methods based on generative adversarial networks (GANs) and semi-supervised models for knowledge transfer, thus improving accuracy and privacy.
- PATE-GAN [11] uses GAN's capabilities to generate synthetic data based on real data using a modified PATE, allowing it to tightly bound the influence of any individual sample on the model. This approach results in tight differential privacy guarantees and thus improved performance over models with the same guarantees.
- In Papernot et al. [19], PATE is applied to larger-scale learning tasks and real-world datasets. Aggregators were improved to allow the application of PATE to uncurated data; in addition, Laplace noise was replaced with Gaussian noise.
- G-PATE [13] also leverages GANs to produce synthetic datasets with strong privacy guarantee. G-PATE ensures differential privacy in the student generator.
- TrPATE [21] modified the original PATE framework and adopted transfer learning to alleviate PATE's performance degradation problem.

However, none of those approaches were applied to bioclinical data, and we found only a limited number of studies applying PATE to bioclinical data. An example is the work of Fay et al. [8,9]. Their studies applied PATE variations to brain tumor segmentation magnetic resonance imaging, as shown in the following references:

– To reduce the required noise level during the aggregation stage, Fay et al. [8] assessed principal component analysis for dimensionality reduction to map the prediction target onto a low-dimensional latent space theoretically and auto-encoders experimentally on a brain tumor dataset. Their study used Gaussian noise in the aggregation stage.

– Autoencoder-based PATE [9] is a PATE variant that builds low-dimensional representations of segmentation masks that the student can obtain through low-sensitivity queries to the private aggregator. This approach achieves a higher Dice coefficient (segmentation quality) for the same privacy guarantee on a brain tumor segmentation dataset.

To our knowledge, at the time of this study, there are no published studies of PATE or any of the PATE variants that address performance scores for the minor classes that have also been applied to bioclinical data on high-performance computers. To help solve these issues, we present the BATE approach, which uses bagging in HPC instead of the data partitioning implemented in the PATE framework. Thus, it generates performance scores for the minor classes at the same time that it includes differential privacy.

The main contributions of our work to the problem we are exploring are the following: 1. We proposed the use of BATE to enhance the PATE differential privacy approach with the use of bagging. 2. We applied BATE to bioclinical data. The results of our study show improvements in those classes suffering from severe class imbalance. 3. The study was performed on a high-performance computer.

In the following section we present the datasets employed in this study and the implementation approach.

3   Methods

3.1   Datasets

The dataset for this study consists of unstructured text in pathology reports from four cancer registries: the Louisiana Tumor Registry, Kentucky Cancer Registry, Utah Cancer Registry, and New Jersey State Cancer Registry. These registries contribute to the National Cancer Institute's SEER program. The study was executed in accordance with the institutional review board protocol DOE000152.

Certified tumor registrars manually coded the ground truth labels associated with each unique case based on free text from the corresponding pathology reports, according to the SEER program coding and staging manual. We consulted the International Classification of Diseases for Oncology, Third Edition, coding convention for labeling the cases. We extracted the following six data fields from the cancer reports: cancer site (70 classes), subsite (320 classes), laterality (7 classes), histology (571 classes), behavior (4 classes), and tumor grade (9 classes). Note that the dataset has a severe class imbalance among class labels (e.g., C50: 242,427 cases, C39: 6 cases), and some labels have few training samples available (e.g., C630: 2 cases, C764: 3 cases), mainly because of the low prevalence of rare cancers.

We chose reports with specimens collected in or after 2017 as testing data and specimens collected in or before 2016 as training data. We randomly selected and reserved 10% of the training data for validation of the model training. Also, we considered only cases for there was less than a 10-day difference between the date of diagnosis and either the specimen collection date or the date of surgery. The 10-day time difference was determined based on an analysis of the pathology report submissions. The vast majority of reports and addenda fell within that time frame. Table 1 lists the number of pathology reports from the four registries. Note that we renamed the registries in the table for security purposes. Note also that, in each registry, there are around 50,000 words in the vocabulary that appeared across the registries.

In this paper, we designed a study in which we selected one registry as a student institution and developed an information extraction DL model with the training data from the other three registries regarding teacher institutions. We repeated this training procedure four times, once per each registry as a student.

Table 1. Number of training and testing cases from the four cancer registries, number of vocabularies in the corpus, and the number of unique words only appearing in the registry. We renamed the registries for security purposes.

Cancer registry	1	2	3	4
Train	147,191	91,820	243,475	259,699
Test	1,554	21,411	58,049	49,433
# Words	479,570	79,959	189,037	247,555
# Unique Words	428,957	35,332	125,787	187,079

3.2   Multi-task Convolutional Neural Networks

We chose the MT-CNN [4] as our DL model for information extraction from cancer pathology reports. It is an extension of the CNN for sentence classification [12,20]. The model consists of three parts: word embedding, one-dimensional convolution, and a task-specific, fully connected layer.

Word embedding is a learned representation of terms to map a set of words onto vectors of numerical values that have the same semantic meaning and have similar observations. A security vulnerability in the word embedding layer is that we can hypothesize that the disease types and the patients' personal information may be clustered together in the vector space.

The convolution layer has a series of one-dimensional convolution filters that have latent representations to capture the features from the word vectors of documents. The algorithm determines the optimal features by itself. However, in the overfitting instances, feature learners may attempt to extract personal identities and sensitive information and become vulnerable to purposeful adversarial attacks.

3.3 Bootstrap Aggregation of Teacher Ensembles Algorithm

In AI and ML, there are two types of information the ML models can observe from the data corpus. One is "common information" that contains concepts and ideas that can articulate data characteristics and their class association. The other is "private information" that is specific to the individual cases and typically should not be contributed to the classification and ML process. However, in certain circumstances, some pieces of private information can be included in the decision process; we refer to the inclusion of such data as "overfitting." The main idea of the PATE algorithm is to divide one training corpus into several subsets that are disjointed from one another and to develop multiple teacher models. The choice of disjoint sets prevents private information from influencing the decision, thus preventing exposure of the identities of individual data subjects in the sensitive data. However, the disjoint data splitting in PATE may cause a considerable performance decrease for decisions in underrepresented classes.

Bootstrap Aggregation. We propose to apply bagging, which is the technique that we do training with multiple models with many sampled data with replacement, thus improving stability and accuracy while helping to avoid overfitting of the data. Both (disjoint) sampling and bagging prevent the extraction of private information from the data; the latter approach maintains or improves performance in classifying minor classes. One drawback is that bagging increases the computational demands for training many models with multiples of the data [22].

Additive Laplace Noise. We added Laplace noise to the teachers' aggregation of predictions, perturbing the counts, and formed a single prediction, which is also known as the "noisymax mechanism" [6]. The purpose of this procedure is to prevent a single outlier from driving decisions when two output classes receive an equal number of votes from the teachers, which could result in the exposure of private information. Additive random noise will not change the decision if it is obvious, thus receives majority votes. Adding a larger scale of noise to the decisions might increase the privacy budget, but it would considerably degrade the overall task performance.

Student Model. Even if the teacher models were trained in a privacy-preserving manner, releasing the models directly to other parties and institutions would present a potential risk of leaking private information because there is a finite privacy budget in the model. Moreover, in cases of natural language processing models, exposure of vocabularies may give a hint to an adversary. Instead, a student institution provides a pilot dataset, and the teacher models derive decisions from the dataset. The student institutions develop their own models based on the pilot dataset with the teacher's annotations.

3.4   Study Design

We designed a study with four participating cancer registries, simulating a situation in which each registry learns from the other three registries. We set up four scenarios as follows:

Scenario 1: No Collaboration. There was no interaction or communication among the cancer registries. Each registry developed its own DL model based on its data and manual annotation. This was the most secure and privacy-preserving method of development, but each institution had to spend effort on it.

Scenario 2: No Privacy-Preserving Collaboration. One institution received a model developed by the data collected from the other three institutions. There was no preparation for an adversary attack on privacy or leaking of personal identity. An institution did not have to spend effort to develop its models, and there was the possibility of a performance boost to some extent because of the abundance of training samples from other institutes.

Scenario 3: PATE. We followed the PATE algorithm: we made 20 disjoint subsamples from the training dataset collected from the three institutes, developed 20 DL models (teacher), and developed one student model trained by the pilot dataset and annotations from 20 teachers. A considerable performance decrease was expected, especially on subsite and histology classification tasks, because those tasks contained several underrepresented class labels.

Scenario 4: BATE. We trained 200 bootstrap sampled datasets and developed 200 teachers. The student model was trained by the pilot dataset with annotations from the 200 teachers. This was the most computationally expensive method of all the scenarios. For the PATE and BATE algorithms in this study, we chose one cancer registry as a student institution and regarded the training set of the registry as the pilot dataset. The student model was trained not by the gold standard of the training set, but by the teachers.

4   Results

We ran experiments in extracting information from cancer pathology reports provided by the four SEER cancer registries, based on the four scenarios described in the previous sections. We extracted the following six properties: primary cancer site, subsite, laterality, histology, behavior, and grade. We performed parallel training and validation of the DL models on the Summit supercomputer operated by the Oak Ridge Leadership Computing Facility (OLCF). The codes were implemented with the Keras and TensorFlow [1] backend available in the IBM Watson ML packages. Since the datasets had many class labels and some had severe class imbalances, we adopted micro and macro-averaged F1 scores as performance metrics. The results are listed in Table 2.

Table 2. Information extraction task performance in micro and macro-averaged F1 metrics for each registry as a student institution and the average from all four registries. S1 (Scenario 1): no collaboration, S2 (Scenario 2): no privacy preservation, S3 (Scenario 3): PATE algorithm, and S4 (Scenario 4): BATE algorithm. λ is the scale of the additive Laplace noise to the aggregated decisions from the teachers.

	Reg.	S1	S2	S3			S4		
λ				0	0.05	0.1	0	0.05	0.1
Site									
Micro F1	1	0.9344	0.9286	0.9279	0.9292	0.9241	0.9324	0.9356	0.9228
	2	0.9287	0.9278	0.9232	0.9218	0.9203	0.9254	0.9264	0.9225
	3	0.9247	0.9202	0.9187	0.9171	0.9130	0.9225	0.9229	0.9215
	4	0.9248	0.9222	0.9145	0.9144	0.9093	0.9238	0.9215	0.9186
	Average	0.9281	0.9247	0.9211	0.9206	0.9167	0.9260	0.9266	0.9213
Macro F1	1	0.6173	0.6491	0.6186	0.6201	0.5892	0.5959	0.6020	0.5934
	2	0.6244	0.6473	0.5653	0.5641	0.5508	0.6209	0.6257	0.6024
	3	0.6424	0.6704	0.5645	0.5544	0.5190	0.6338	0.6373	0.6163
	4	0.6545	0.6355	0.5473	0.5456	0.5080	0.6374	0.6417	0.6102
	Average	**0.6346**	**0.6506**	**0.5739**	**0.5710**	**0.5418**	**0.6220**	**0.6267**	**0.6056**
Subsite									
Micro F1	1	0.5978	0.5927	0.5882	0.5759	0.5592	0.6004	0.6010	0.5766
	2	0.6578	0.6513	0.6439	0.6431	0.6355	0.6634	0.6637	0.6492
	3	0.6435	0.6347	0.6153	0.6135	0.6003	0.6530	0.6543	0.6444
	4	0.6467	0.6490	0.6310	0.6302	0.6231	0.6548	0.6531	0.6429
	Average	**0.6365**	**0.6319**	**0.6196**	**0.6157**	**0.6045**	**0.6429**	**0.6430**	**0.6283**
Macro F1	1	0.3794	0.3573	0.3094	0.2963	0.2907	0.3170	0.3207	0.3068
	2	0.3087	0.3143	0.2391	0.2357	0.2107	0.2953	0.2975	0.2596
	3	0.2771	0.2956	0.2148	0.2037	0.1849	0.2870	0.2818	0.2515
	4	0.3060	0.3029	0.2203	0.2127	0.1874	0.2940	0.2891	0.2574
	Average	**0.3178**	**0.3175**	**0.2459**	**0.2371**	**0.2184**	**0.2983**	**0.2973**	**0.2688**
Laterality									
Micro F1	1	0.9157	0.9028	0.9125	0.9138	0.9151	0.9208	0.9176	0.9118
	2	0.9130	0.9021	0.9029	0.9029	0.8977	0.9030	0.9038	0.9006
	3	0.9036	0.9003	0.9048	0.9054	0.9030	0.9048	0.9041	0.9007
	4	0.9023	0.8982	0.9012	0.9005	0.8990	0.9045	0.9042	0.8983
	Average	**0.9086**	**0.9009**	**0.9053**	**0.9056**	**0.9037**	**0.9083**	**0.9074**	**0.9029**
Macro F1	1	0.5920	0.4693	0.5536	0.5544	0.5592	0.4777	0.4726	0.5092
	2	0.5221	0.5179	0.5097	0.5155	0.5057	0.5201	0.5295	0.5116
	3	0.5296	0.5124	0.5004	0.5066	0.4915	0.5210	0.5231	0.5047
	4	0.5265	0.5141	0.5086	0.5095	0.5039	0.5167	0.5173	0.5039
	Average	**0.5426**	**0.5034**	**0.5180**	**0.5215**	**0.5151**	**0.5089**	**0.5106**	**0.5074**
Histology									
Micro F1	1	0.7252	0.7207	0.7130	0.7072	0.6963	0.7291	0.7246	0.7079
	2	0.7469	0.7414	0.7310	0.7304	0.7217	0.7411	0.7394	0.7300

(continued)

Table 2. (*continued*)

	Reg.	S1	S2	S3			S4		
λ				0	0.05	0.1	0	0.05	0.1
	3	0.7546	0.7453	0.7469	0.7462	0.7425	0.7607	0.7613	0.7536
	4	0.7803	0.7756	0.7674	0.7639	0.7580	0.7755	0.7783	0.7674
	Average	**0.7518**	**0.7457**	**0.7396**	**0.7369**	**0.7296**	**0.7516**	**0.7509**	**0.7397**
Macro F1	1	0.4004	0.3906	0.3047	0.3224	0.2544	0.3609	0.3732	0.3472
	2	0.3540	0.3444	0.2245	0.2193	0.1873	0.3096	0.3120	0.2605
	3	0.3239	0.3164	0.1998	0.1859	0.1532	0.3009	0.2981	0.2517
	4	0.3275	0.3276	0.2041	0.1993	0.1452	0.3142	0.3036	0.2551
	Average	**0.3515**	**0.3448**	**0.2333**	**0.2317**	**0.1850**	**0.3214**	**0.3217**	**0.2786**
Behavior									
Micro F1	1	0.9704	0.9743	0.9698	0.9665	0.9659	0.9736	0.9710	0.9646
	2	0.9654	0.9585	0.9595	0.9575	0.9560	0.9598	0.9617	0.9570
	3	0.9671	0.9665	0.9684	0.9678	0.9644	0.9696	0.9688	0.9672
	4	0.9731	0.9670	0.9693	0.9680	0.9655	0.9709	0.9703	0.9680
	Average	**0.9690**	**0.9666**	**0.9667**	**0.9650**	**0.9630**	**0.9685**	**0.9680**	**0.9642**
Macro F1	1	0.8159	0.8595	0.7201	0.7533	0.6460	0.8730	0.8076	0.8373
	2	0.9038	0.8664	0.8094	0.8511	0.8073	0.8554	0.8654	0.8507
	3	0.8133	0.8316	0.7378	0.7393	0.7057	0.8363	0.8151	0.7581
	4	0.8207	0.8389	0.7674	0.7542	0.7265	0.8417	0.8267	0.7893
	Average	**0.8384**	**0.8491**	**0.7587**	**0.7745**	**0.7214**	**0.8516**	**0.8287**	**0.8089**
Grade									
Micro F1	1	0.7259	0.6731	0.6763	0.6692	0.6744	0.6737	0.6660	0.6577
	2	0.7807	0.7732	0.7728	0.7651	0.7673	0.7817	0.7801	0.7752
	3	0.7255	0.7115	0.7196	0.7210	0.7112	0.7279	0.7293	0.7204
	4	0.7587	0.7461	0.7564	0.7552	0.7498	0.7571	0.7586	0.7503
	Average	**0.7477**	**0.7260**	**0.7313**	**0.7276**	**0.7257**	**0.7351**	**0.7335**	**0.7259**
Macro F1	1	0.7364	0.7090	0.7055	0.7046	0.6989	0.7200	0.6910	0.6871
	2	0.6011	0.6297	0.5885	0.5812	0.5803	0.6067	0.6327	0.5968
	3	0.6503	0.6083	0.5631	0.5697	0.5558	0.6220	0.6269	0.5771
	4	0.7716	0.6961	0.6720	0.6728	0.6664	0.6772	0.7453	0.6703
	Average	**0.6898**	**0.6608**	**0.6323**	**0.6321**	**0.6253**	**0.6565**	**0.6740**	**0.6328**

We observed that the F1 scores between S1 (no collaboration) and S2 (transfer knowledge without privacy preservation) were very close. That finding was the confirmation that we could achieve a similar level of task performance by developing a model with the other registries' data and testing it with the student registry. It implies that our study design is legitimate.

Based on the comparisons of F1 scores between S2 and S3 or S4, we observed a certain level of performance decrease if we applied privacy-preserving algorithms; that finding is confirmed by other studies [2] showing that there is a trade-off between accuracy and privacy. However, we observed more degradation of performance from applying the PATE algorithm (S3) than from applying the BATE (S4). That was especially true for the macro-averaged F1 scores of subsite (S3 averaged macro-F1: 0.2459, S4: 0.2983) and histology (S3: 0.2333, S4: 0.3214), two tasks that have many underrepresented class labels. It was a clear demonstration that BATE performance was superior to PATE performance.

The results also supported the findings of other studies that adding more noise to the decision may increase the privacy budget but decrease the classification accuracy [16]. Both the S3 and S4 scenarios showed that increasing the scale parameter of the additive Laplace noise lowered the classification accuracy scores. Those findings were more clear for the macro-F1 scores of the subsite and histology tasks. Also, we observed that BATE performance was superior to PATE performance for the subsite (S3: 0.2184, S4: 0.2688) and histology (S3: 0.1850, S4: 0.2786) tasks.

5 Discussion

Threats to data privacy in AI and ML/DL are incurred as a result of the nature of the design: ML/DL models are trained without having domain knowledge but find the best feature representations that can maximize the task performance. During the training, the algorithm may learn too precisely from the examples and may attempt to extract personal and sensitive information. The state-of-art differential privacy algorithms are designed primarily to avoid such incidents so that the few marginal training samples dominate decisions. We suggested the BATE algorithm, in which we adopted the advantages of PATE so that we could isolate the vocabulary sets of the student institutions from the teacher institutions and limit the access of teacher models to secure privacy. Also, with the BATE model, we maintained the accuracy scores of the underrepresented classes of the training samples.

Information extraction from cancer pathology reports was our model example. There were many class labels in the dataset, including those of rare cancer types, which resulted in severe class imbalances and underrepresentation of training examples. We demonstrated that BATE performance was superior to PATE performance, especially for those difficult problems. We also showed that, with BATE, the privacy-preserving training and transfer of knowledge from the teacher institutions to the student institutions maintained the clinical task performance.

The study's limitation is that we examined the effects of the BATE algorithm qualitatively but did not quantify the threat of privacy and security attacks from the adversary. The results suggested that we need to design a follow-up study to confirm the validity and security of the privacy-preserving knowledge transfer.

Acknowledgement. This research was supported by the Exascale Computing Project (17-SC-20-SC), a collaborative effort of the US Department of Energy (DOE) Office of Science and the National Nuclear Security Administration. This work has been supported in part by the Joint Design of Advanced Computing Solutions for Cancer (JDACS4C) program established by DOE and the National Cancer Institute of the

National Institutes of Health. This work was performed under the auspices of DOE by Argonne National Laboratory under Contract DE-AC02-06-CH11357, Lawrence Livermore National Laboratory under Contract DE-AC52-07NA27344, Los Alamos National Laboratory under Contract DE-AC5206NA25396, and ORNL under Contract DE-AC05-00OR22725.

KCR data were collected with funding from NCI Surveillance, Epidemiology and End Results (SEER) Program (HHSN261201800013I), the CDC National Program of Cancer Registries (NPCR) (U58DP00003907) and the Commonwealth of Kentucky.

LTR data were collected using funding from NCI and the Surveillance, Epidemiology and End Results (SEER) Program (HHSN261201800007I), the CDC's National Program of Cancer Registries (NPCR) (NU58DP006332-02-00) as well as the State of Louisiana.

NJSCR data were collected using funding from NCI and the Surveillance, Epidemiology and End Results (SEER) Program (HHSN261201300021I), the CDC's National Program of Cancer Registries (NPCR) (NU58DP006279-02-00) as well as the State of New Jersey and the Rutgers Cancer Institute of New Jersey.

The Utah Cancer Registry is funded by the National Cancer Institute's SEER Program, Contract No. HHSN261201800016I, and the US Centers for Disease Control and Prevention's National Program of Cancer Registries, Cooperative Agreement No. NU58DP0063200, with additional support from the University of Utah and Huntsman Cancer Foundation.

The study was supported by the Laboratory Directed Research and Development (LDRD) program of Oak Ridge National Laboratory, under LDRD project No. 9831.

This research used resources of the Oak Ridge Leadership Computing Facility at ORNL, which is supported by the DOE Office of Science under Contract No. DE-AC05-00OR22725.

References

1. Abadi, M., et al.: Tensorflow: a system for large-scale machine learning. In: 12th {USENIX} Symposium on Operating Systems Design and Implementation ({OSDI} 2016), pp. 265–283 (2016)

2. Abadi, M., et al.: Deep learning with differential privacy. In: Proceedings of the 2016 ACM SIGSAC Conference on Computer and Communications Security, pp. 308–318 (2016)

3. Abadi, M., et al.: On the protection of private information in machine learning systems: two recent approches. In: 2017 IEEE 30th Computer Security Foundations Symposium (CSF), pp. 1–6. IEEE (2017)

4. Alawad, M., et al.: Automatic extraction of cancer registry reportable information from free-text pathology reports using multitask convolutional neural networks. J. Am. Med. Inform. Assoc. **27**(1), 89–98 (2020)

5. Dwork, C.: Differential privacy: a survey of results. In: Agrawal, M., Du, D., Duan, Z., Li, A. (eds.) TAMC 2008. LNCS, vol. 4978, pp. 1–19. Springer, Heidelberg (2008). https://doi.org/10.1007/978-3-540-79228-4_1

6. Dwork, C., McSherry, F., Nissim, K., Smith, A.: Calibrating noise to sensitivity in private data analysis. In: Halevi, S., Rabin, T. (eds.) TCC 2006. LNCS, vol. 3876, pp. 265–284. Springer, Heidelberg (2006). https://doi.org/10.1007/11681878_14

7. Dwork, C., Roth, A., et al.: The algorithmic foundations of differential privacy. Found. Trends® Theoret. Comput. Sci. **9**(3–4), 211–407 (2014)

8. Fay, D., Sjölund, J., Oechtering, T.J.: Private learning for high-dimensional targets with pate (2020)
9. Fay, D., Sjölund, J., Oechtering, T.J.: Decentralized differentially private segmentation with pate. arXiv preprint arXiv:2004.06567 (2020)
10. Fung, B.C., Wang, K., Chen, R., Yu, P.S.: Privacy-preserving data publishing: a survey of recent developments. ACM Comput. Surv. (CSUR) **42**(4), 1–53 (2010)
11. Jordon, J., Yoon, J., van der Schaar, M.: Pate-GAN: generating synthetic data with differential privacy guarantees (2018)
12. Kim, Y.: Convolutional neural networks for sentence classification. arXiv preprint arXiv:1408.5882 (2014)
13. Long, Y., Lin, S., Yang, Z., Gunter, C.A., Li, B.: Scalable differentially private generative student model via pate. arXiv preprint arXiv:1906.09338 (2019)
14. McMahan, H.B., et al.: A general approach to adding differential privacy to iterative training procedures. arXiv preprint arXiv:1812.06210 (2018)
15. Papernot, N., Abadi, M., Erlingsson, Ú., Goodfellow, I., Talwar, K.: Machine learning with privacy by knowledge aggregation and transfer
16. Papernot, N., Abadi, M., Erlingsson, U., Goodfellow, I., Talwar, K.: Semi-supervised knowledge transfer for deep learning from private training data. arXiv preprint arXiv:1610.05755 (2016)
17. Papernot, N., McDaniel, P., Sinha, A., Wellman, M.: Towards the science of security and privacy in machine learning. arXiv preprint arXiv:1611.03814 (2016)
18. Papernot, N., McDaniel, P., Sinha, A., Wellman, M.P.: Sok: security and privacy in machine learning. In: 2018 IEEE European Symposium on Security and Privacy (EuroS&P), pp. 399–414. IEEE (2018)
19. Papernot, N., Song, S., Mironov, I., Raghunathan, A., Talwar, K., Erlingsson, Ú.: Scalable private learning with pate. arXiv preprint arXiv:1802.08908 (2018)
20. Qiu, J.X., Yoon, H.J., Fearn, P.A., Tourassi, G.D.: Deep learning for automated extraction of primary sites from cancer pathology reports. IEEE J. Biomed. Health Inform. **22**(1), 244–251 (2017)
21. Wang, L., Zheng, J., Cao, Y., Wang, H.: Enhance pate on complex tasks with knowledge transferred from non-private data. IEEE Access **7**, 50081–50094 (2019)
22. Yoon, H.J., et al.: Accelerated training of bootstrap aggregation-based deep information extraction systems from cancer pathology reports- manuscript submitted for publication

An Intelligent and Efficient Rehabilitation Status Evaluation Method: A Case Study on Stroke Patients

Yao Tong[1,2] ⓘ, Hang Yan[1], Xin Li[3], Gang Chen[1](✉), and Zhenxiang Zhang[1](✉)

[1] Zhengzhou University, Zhengzhou 450000, Henan, China
yaotong@uw.edu, chengang@zzu.edu.cn, zhangzx6666@126.com
[2] University of Washington, Seattle, WA 98105, USA
[3] Hohai University, Nanjing 211000, Jiangsu, China

Abstract. Chronic patients' care encounters challenges, including high cost, lack of professionals, and insufficient rehabilitation state evaluation. Computer-supported cooperative work (CSCW), is capable of alleviating these issues, as it allows healthcare physicians (HCP) to quantify the workload and thus to enhance rehabilitation care quality. This study aims to design a deep learning algorithm Pose-AMGRU, a deep learning-based pose recognition algorithm combining Pose-Attention Mechanism and Gated Recurrent Unit (GRU), to monitor the human pose of rehabilitating patients efficiently. It gives instructions for HCP. To further substantiate the acceptance of our computer-supported method, we develop a multi-fusion theoretical model to determine factors that may influence the acceptance of HCP and verify the usefulness of the method above. Experiment results show Pose-AMGRU achieves an accuracy of 98.61% in the KTH dataset and 100% in the rehabilitation action dataset, which outperforms other algorithms. The efficiency running speed of Pose-AMGRU on the GTX1060 graphics card is up to 14.75FPS, which adapts to the home rehabilitation scene. As to acceptance evaluation, we verified the positive relationship between the computer-supported method and acceptance, and our model presents decent generalizability of stroke patients' care at the Second Affiliated Hospital of Zhengzhou University.

Keywords: Rehabilitation status evaluation · Deep learning · Technical acceptance

1 Introduction

Chronic diseases, such as diabetes, stroke, and heart disease, etc. are the major diseases that deteriorate the life quality of the elderly and bring our health care system high cost $199 billion per year. Efficiency and quality of caring are essential to the health of a

© Springer Nature Switzerland AG 2021
V. Gadepally et al. (Eds.): Poly 2020/DMAH 2020, LNCS 12633, pp. 100–119, 2021.
https://doi.org/10.1007/978-3-030-71055-2_10

nation [1–4]. Chronic illness has various long-term sequelae that need rehabilitation, and patients with chronic disease usually choose to go to hospital or community medical institutions for follow-up rehabilitation after discharge [5]. In the case of stroke patients, stroke after nerve trauma is one of the leading causes of long-term disability in adults. Up to 30% of stroke survivors experience minimal exercise recovery and rely on assistance in managing their daily activities [6]. As population aging and the trend of stroke increasing among younger people [7], there are still problems with efficiency and quality in rehabilitation care management. Delivering efficient and high-quality healthcare is complicated [8–12] since most of HCP's time is preoccupied with hunting for supplies, tracking down medications, filling out paperwork at the nursing station, and looking for missing test results [13].

Computer-supported cooperative work (CSCW) is defined that in the environment supported by computer technology, groupware cooperatively works to accomplish a common task [14]. The previous researches were focused on the behaviors of individuals at scale [15, 16]. Therefore, this kind of computer-supported method has limited generalization ability when facing a large number of users. Recently, there are many Computer-supported approaches to enhance efficiency or quality of healthcare service, such as a single application of computer algorithms to optimize and support the workflow of HCP or a mobile service app to help patients realize a better self-management [17]. However, there are few systematic theoretical studies on computer-supported healthcare services and complete process framework studies. To address the problem mentioned above, in this paper, we 1) propose the human rehabilitation movement recognition algorithm, Pose-AMGRU, which helps HCP to monitor stroke patients' state of recovery, and 2) construct a theoretical model of multi-model fusion to discuss the acceptance and adoption of our computer-supported method. Previous studies on the acceptability of relevant technologies are mostly aimed at patients [18, 19]. However, HCP is a group that needs more of such techniques because the efficiency and quality of HCP rehabilitation care management have a more direct impact on patients' rehabilitation and self-management [20].

2   Human Rehabilitation Movement Recognition Algorithm

As shown in Fig. 1, our human rehabilitation movement recognition algorithm is mainly composed of human posture estimation, preprocessing, feature extraction, and classification network, respectively.

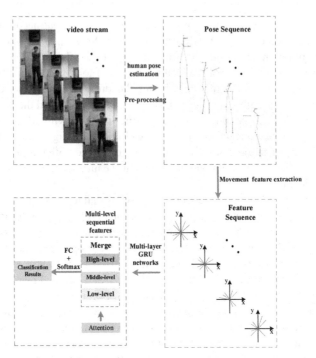

Fig. 1. Framework of pose recognization Pose-AMGRU.

2.1 Data Set

In this paper, a standardized rehabilitation action of stroke patients is referred to [21], and a rehabilitation action dataset is built under the guidance of professional physicians. Meanwhile, we chose an open data set KTH [22] to evaluate the performance of the algorithm. A data instance of KTH is shown in Fig. 2(1).

KTH is a widely-used dataset in the field of motion recognition, consisting of six movements recorded by 25 volunteers, including walking, jogging, running, clapping, waving, and boxing. The dataset consists of 599 videos, which can be subdivided into 2,391 action segments. The videos in KTH, which contain the whole target human body, is a single-person scene that can detect the complete human posture node. Therefore, we choose this dataset to do the comparison experiment.

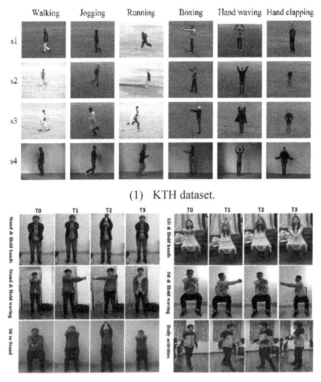

(1)   KTH dataset.

(2)   Rehabilitation action dataset.

Fig. 2.   Data set.

As shown in Fig. 2(2), our rehabilitation action dataset is collected by ten volunteers in 6 different environments, including a total of 2075 video segments of 6 types of behaviors, with changes in light, background, and distance. The behavior types are divided into five kinds of rehabilitation actions and one kind of daily activity. Daily activities include strolling, stretching, sitting still, and standing still. The video is 15 frames per second and lasts between 7 and 15 s.

2.2   Pose Estimate from Videos

We utilize OpenPose [23] to estimate human pose and to detect skeleton joints from videos. OpenPose is a real-time pose estimation model based on the top-down approach and deep learning, which can detect a human face, trunk, limbs, hand bone points, and maintain speed advantage in multiple scenes, as shown in Fig. 3.

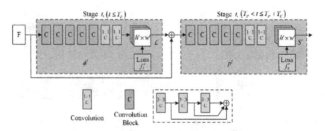

Fig. 3. OpenPose network architecture.

Latest OpenPose network architecture adopts the method of multi-stage prediction. The framework of OpenPose takes the VGG-19 model as the foundation of the top 10 layer network, transforming the input image into feature F, through multiple convolution neural network regression $L(p)$ and $S(p)$. $L(p)$ means PAFs (Part Affinity Fields), which describes the key points in the skeleton $S(p)$, represents the confidence of the joint and describes the position information of the joint. The model divides the prediction process into two different stages, such that the first T_p stage predicts the affinity vector field L^t, and the last T_c stage predicts confidence S^t.

$$L^1 = \phi^1(F), t = 1 \tag{1}$$

$$L^t = \phi^t(F, L^{t-1}), 2 \leq t \leq T_P \tag{2}$$

After iterations of T_p, the process is repeated for the confidence maps detection, starting from the most updated part affinity field prediction.

$$S^{T_P} = \phi^t(F, L^{T_P}), t = T_P \tag{3}$$

$$S^t = \phi^t(F, L^{T_P}, S^{t-1}), T_P < t \leq T_P + T_C \tag{4}$$

At each stage, the results of the previous stage are fused with the original features to preserve both the lower-layer and higher-layer features of the image. After predicting the position and affinity vectors of the joint nodes, we leverage the Hungarian algorithm to perform optimal binary matching for adjacent joint nodes to obtain the pose information belonging to the same body.

2.3 Classification Network

In this paper, we design Pose-AMGRU, an improved GRU classification network [24] to recognize different layers of human pose and to process images. This classification network is based on convolutional neural networks fused with multilevel spatial features, such as SSD [25], DenseNet [26]. At the same time, we combine the attention mechanism to enhance the salience of features. The designed classification network is shown in Fig. 4.

The input of the model is 26 action features extracted from each frame, and the time step size is T. MK is a Masking layer used for supporting variable-length sequences. Time steps with eigenvalues all of 0 are ignored in GRU recursive computation. In the Normalization layer, to speed up the convergence process during training, we implement batch normalization to the learnable parameters with Gaussian distribution of a mean value of 0 and a variance of 1. The number of network neurons in each layer of our three-layer stacked GRU cell network is 64, and the output state h of the underlying cell network at all times is transferred to the next layer. Leveraging attention mechanism to calculate the attention weight of output features of each time step, and the space-time characteristics of each layer are obtained by the weighted sum of output features and attention weight at different moments.

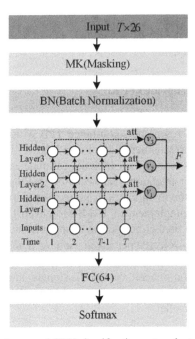

Fig. 4. Improved GRU classification network structure.

3   Construction of Theoretical Model

In this part, we utilize multi-fusion models to construct an acceptance theory, which can verify our computer-supported method is useful and efficient for healthcare management.

3.1   Technology Acceptance Model (TAM)

TAM is a classical model that applies contexts for predicting and evaluating user acceptance of information technology [27]. The model is mainly composed of four levels, which are external stimulus, cognitive, affective and behavioral response, and the validity of the model for considering the users' acceptance of information technology is verified. Many subsequent researches leveraged TAM in their studies [28–31] to explain physicians' decisions to accept telemedicine technology in the healthcare context. Moreover, with the popularity of smartphone applications, Yangil [32] utilized TAM to investigate human motivations affecting an adoption decision for the smartphone among medical doctors and nurses, and they found that HCP's attitude toward the use of smartphone technology is positive. These findings lead us to hypothesize the following:

H1: *Perceived ease of use* is positively associated with HCP's acceptance of computer-supported healthcare delivery.
H2: *Perceived usefulness* is positively associated with HCP's acceptance of computer-supported healthcare delivery.

3.2   Theory of Planned Behavior (TPB)

Realizing the limitation of TAM in explaining differences in subjectively perceived ease of use and usefulness [33], researchers combined TPB with TAM to achieve a better interpretation effect. Many researchers used the TPB to verify the behavior change of HCP or patients. TPB exists in five parts, including attitude, subjective norm, perceived behavioral control, behavior intention, and behavior, and these five variables, and all these components point to the intention of using the technology eventually. Therefore, TPB can be an excellent complement to TAM. Moreover, Patrick [34] leveraged the decomposed TPB model to predict behavioral changes in their patients to utilize TPB more flexible, such that they selected perceived usefulness and perceived ease of use as the mediating variables. Gaston [35] documented the cognitive factors most consistently associated with the prediction of healthcare professionals' intention and behaviors, and they found TPB appears to be an appropriate theory to predict behavior. Therefore, the following assumptions are made:

H3: *Subjective norm* is positively associated with HCP's acceptance of computer-supported healthcare delivery.
H4: *Intention to use* is positively associated with HCP's acceptance of computer-supported healthcare delivery.

3.3 Unified Theory of Acceptance and Use of Technology (UTAUT)

Venkatesh [36] made a statistical difference analysis of previous studies and designed a unified theory of acceptance and use of technology model (UTAUT). UTAUT was formulated with four core determinants of intention and usage, and up to four moderators of key relationships (Performance Expectancy (PE), Effort Expectancy (EE), Social Influence (SI), and Facilitating Conditions (FC). Performance and effort expectancy are the first shape of their belief in the design of the UTAUT theory. Moreover, simple to use is also emphasized in UTAUT, since if HCP gets sufficient technical support to perceive the ease of use of new technology, HCP's acceptance, and motivation will be improved. Similarly, if HCP can get adequate technical support to let them perceive the ease of use of new technology, HCP's acceptance and motivation can be improved. Therefore, we propose the following hypotheses:

H5a: *Expectancy* is positively associated with HCP's perceived ease of use of computer-supported healthcare delivery.
H5b: *Expectancy* is positively associated with HCP's perceived usefulness of computer-supported healthcare delivery.
H6: *Facility conditions* are positively associated with HCP's perceived ease of use of computer-supported healthcare delivery.

Moreover, UTAUT can be utilized with some additional contextual constructs that integrate specific elements of the field of use, such as social factors or personal experience [37, 38]. As shown in Cimperman's study [39], support from managers, peers, and colleagues, or other relevant people can also affect HCP's perceived usefulness of new technology. Therefore, we propose the following hypotheses:

H7: *Social influence* is positively associated with the perceived usefulness of computer-supported healthcare delivery.

However, many previous researches on the acceptance of technology focused on the psychological and behavioral factors of users. Still, there is a limited in-depth discussion of relevant technical factors. Hence, our goal is to propose a technic acceptance theoretical model on real-time human pose estimate technology that can be applied in rehabilitation care management. We select the CMT and SEIPS model fused into our novel model. After the primary interview with HCP in our hospital, we propose the following hypotheses:

H8: *Technical quality* is positively associated with HCP's acceptance of computer-supported healthcare delivery.

3.4 Care Management Technology (CMT)

CMT [40] is used by community health workers (CHWs) and care managers (CMs) working collaboratively to improve risk factor control among recent stroke survivors. It has been proved that CMT can enhance the effectiveness of the CHWs team. CMT is readily accepted by HCP and their managers, as it can support the electronic collection of clinical evaluation data, provide decision support, and obtain patients' risk factor values remotely. But as the researchers said, one weakness of CMT is the slow rate of reaction, that the gap between the electronic technic and HCP who is not familiar with the technical results to the delayed response. Therefore, in our research, we design a real-time computer-supported method to enhance both HCP and patients' experience. Moreover, if technology and HCP's work are well-compactible to each other, HCP's willingness to accept can be positively affected. Therefore, we decomposed hypotheses into two parts—quality and compatibility:

H8: *Quality* is positively associated with HCP's perceived usefulness of computer-supported healthcare delivery.
H9a: *Compatibility* is positively associated with HCP's perceived ease of use of computer-supported healthcare delivery.
H9b: *Compatibility* quality is positively associated with HCP's perceived usefulness of computer-supported healthcare delivery.

3.5 Systems Engineering Initiative for Patient Safety (SEIPS) Model

SEIPS model is anchored within the industrial engineering subspecialty of human factors [41], as it has particularly embraced three core social factors principles: system orientation, person-centeredness, and design-driven improvements [42]. Carayon [41] leveraged the SEIPS model to support processes and outcomes in the system. The model is proposed based on Donabedian's SPO model [43], which examines the clinical processes and outcomes of care. SEIPS emphasized specific individuals in a healthcare scenario, such as HCP in our research, as the center of the work system, thereby enhance and facilitate the performance of HCP as the center of the work system, thus enhance and facilitate the performance of HCP. Many researches used SEIPS to check work error and care quality or optimize healthcare work management [44]. We thus propose the following hypotheses:

H10: *Safety* is positively associated with HCP's perceived usefulness of computer-supported healthcare delivery.

3.6 The Proposed Theoretical Research Model

Based on the above theoretical framework, we propose a conceptual model (shown in Fig. 5) from multi-models by five different hierarchies corresponding to three different levels of problem analysis, including cognitive, technical, and personal levels. The three cognitive factors are self-efficacy, expectancy (including performance expectancy and effort expectancy), and facility conditions. The two technical factors, which are quality, and compatibility, are measured by three factors, including recordable, real-time, and feedback. The three personal factors are social influence, safety, and intention to use. Furthermore, three factors - expectancy, facility conditions, and compatibility affect HCP's perceived ease of use of the computer-supported method. And five factors - expectancy, compatibility, assistant quality, social influence, and safety affect HCP's perceived usefulness of the computer-supported method. Finally, perceived ease of use and usefulness, self-efficacy, and intention to use are the predictive factors to determine HCP's acceptance of the computer-supported method.

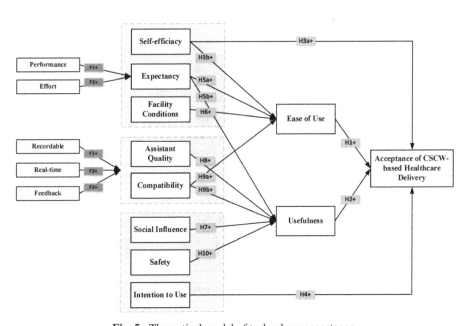

Fig. 5. Theoretical model of technology acceptance.

4   Methods

4.1   Pose-AMGRU Algorithm Application

We implement our Pose-AMGRU algorithm in the rehabilitation departments of two hospitals and two community hospitals and explain how the algorithm works. We use high precision 1080P monocular camera to collect real-time monitoring video streams for online real-time behavior recognition. Smartphones and surveillance cameras are used to collect rehabilitation action data. The critical point information of each skeleton is extracted frame by frame from the video by a multi-human pose estimation method. In the course of the demonstration, the behavior types are divided into three kinds of rehabilitation actions and one kind of regular activities, including upper and lower arm exercises, left and right arm exercises, and sitting up exercises. Regular activities include standing still, sitting still, strolling, stretching, and other daily activities. The resolution of each video is 1280 * 720 or 1920 * 1080, the frame rate is 15 frames per second, and the duration of the video segment is between 8 and 15 s. Based on the traditional method, we have realized the function of real-time online recognition, with an average recognition rate of 90.66%. We can independently deterimine the rehabilitation action of each target in the monitoring flow, and output the position, type, and probability of action in real-time, as shown in Fig. 6.

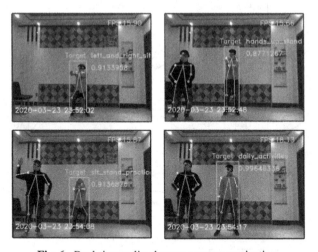

Fig. 6. Real-time online human-pose recognization.

4.2   Conducting a Questionnaire Survey

For our technology acceptance model, we design a questionnaire about influence factors including perceived ease of use (PEOU), perceived usefulness (PU), self-efficiency (SE), expectancy (EXP), facility conditions (FC), compatibility (COM), assistant quality (AQ), social influence (SI), safety (SAF), intention to use (ITU). To ensure the validity of all measures, the measurement items of the latent constructs in the model are developed from previous studies. The detailed items of each construct are listed in Table 2. The concept of technology acceptability was first proposed in the literature [27], in which the authors gave sufficient reasons to prove that perceived ease of use (PEOU) and perceived usefulness (PU) are the major factors positively associated with user's acceptance of new technology, and hence these two variables are widely used in models related acceptance. Our theoretical model takes inspiration from the questionnaires from the literature [45], and the results show excellent reliability. Literatures [34] summarizes factors that affect user's cognition, and they designed path model of TOE–TAM and a decomposed TPB Model based on TAM, which both mentioned that both PEOU and Self efficacy (SE) could affect users' cognition and mentioned one new concept, ITU, which is another factor could affect acceptance of users directly, in our research, we used questionnaires from the literature [46] and [21]. Moreover, the literatures mentioned above indicated that subjective norm and COM could affect PEOU and PU of users. According to the literature [39], behavioral intention to use is influenced by four primary constructs: Performance Expectancy (PE), Effort Expectancy (EE), Social Influence (SI), and Facilitating Conditions (FC). In the subsequent actual investigation, we include all four indicators in our survey due to the mutual complement between subjective norm and SE. We combined the two variables Performance Expectancy (PE), Effort Expectancy (EE), according to the expert consultation. As for PU, apart from considering the cognitive of users, the quality of new technology also needs to be included. Literature [47] summarized a system that could encourage the active and adaptive role of HCP and hence can deliver high-quality healthcare for patients, and the principle of this work system is simply constructed with man-machine cooperation. We revise the questionnaires from literatures [45] and [48], summarize the variable as assistant quality (AQ). After interviewing with HCP, we find that the healthcare management core is person-centered. This is also in line with the concept of human engineering [42], which mentions that the perception of the behavior of people in daily work has a direct impact on the products, equipment, environment, and safety. There were more than four variables used in our questionnaire at first, and after the reliability test, we removed two low-reliability factors and added the action recognition tools to the AQ-related questions. We use Social influence (SI) according to the research [49]. We revised the questionnaires from the literature [50] to design questions about factor Safety (SAF).

　　To verify the reliability and validity of the theoretical model, we design a questionnaire about the model of influence factors on a total of 28 variables. Among the 300 sent questionnaires, we obtain 263 valid recycling copies, and we use the Likert scale, which ranges from 1 (strongly disagree) to 5 (strongly agree), to score each question (Table 1).

Table 1. The questionnaire form.

Construct		Variables	Measurement items	Source
Perceived ease of use (PEOU)		PEOU1	It is easy to operate action recognition tools	[27, 45]
		PEOU2	Learning how to use a new action recognition tools would be easy for me	
		PEOU3	It is easy to monitor patients' statements with action recognition tools	
		PEOU4	Interacting with action recognition tools is often frustrating	
Perceived usefulness (PU)		PU1	Appling action recognition tools in my job would enable me to accomplish tasks more quickly	[33]
		PU2	Appling action recognition tools improve my job performance	
		PU3	Appling action recognition tools would make it easier to do my job	
		PU4	Overall, action recognition tools are useful in my job	
Self-efficacy (SE)		SE1	I can use action recognition tools without much time and energy	[21]
		SE2	I get the best value from using action recognition tools Assistant work	
Expectancy (EXP)	Performance (PE)	PE1	I will increase the quality of the output of my job	[36]

(*continued*)

Table 1. (*continued*)

Construct		Variables	Measurement items	Source
		PE2	My coworkers will perceive me as competent	
	Effort (EE)	EE1	I will spend less time on routine job tasks	
		EE2	I will increase the quantity of output for the same amount of effort	
Facility conditions (FC)		FC1	I believe the guidance will be available to me when deciding whether to use action recognition tools	[39]
		FC2	I believe specific persons (or a group) will be available for assistance with action recognition tools difficulties	
Compatibility (COM)		COM1	Using action recognition tools is compatible with all aspects of my work	[34]
		COM2	Using action recognition tools fits into my work style	
		COM3	Using action recognition tools fits with my service needs	
		COM4	Using action recognition tools does not fit with my practice preferences	
Assistant quality (AQ)		AQ1	Action recognition tools can provide useful information about patients' rehabilitation information	[45, 48]

(*continued*)

Table 1. (*continued*)

Construct		Variables	Measurement items	Source
		AQ2	Using action recognition tools will improve the quality of healthcare services in my city	
Social influence (SI)		SI1	People who are essential in assessing my patient care and management think that I should use action recognition tools	[49]
		SI2	People who influence my behavior would think that I should use action recognition tools	
Safety (SAF)		SAF1	A better self-management can be available with action recognition tools	[50]
		SAF2	Action recognition tools can help me to keep a good nurse-patients relationship	
Intention to use (ITU)		ITU1	Given the opportunity, I would like to use action recognition tools	[21, 46]
		ITU2	I would consider using action recognition tools continuously	

5   Results

5.1   Results on Pose-AMGRU Algorithm

In our dataset, skeleton joints are first extracted from the video and preprocessed, and then compared with various timing relation models based on pose features. The running speed is the total predicted time of all test samples, excluding the calculation time of pose estimation and preprocessing.

We test our model both on the KTH dataset and our dataset. At the same time, we compare our model with four other methods tested on KTH. The experimental result is shown in Table 2 and Table 3. The performance of the cyclic neural network series model is superior to the traditional hidden Markov model, while our model achieves the best recognition results; however, our model requires more computation and hence can affect the real-time performance in practice.

Table 2. Results on the KTH dataset.

Methods	Accuracy (%)
3D CNN [51]	90.20
SVM [52]	94.39
YOLO + CNN & LSTM [53]	96.63
CNN + SVM & KNN [54]	98.15
Our model	98.61

Table 3. Results on the rehabilitation action dataset.

Methods	Accuracy (%)	Time (ms)
HMM	77.52	911
RNN	94.06	673
LSTM	98.39	1128
IndRNN	99.19	709
SRU	98.07	741
Our model	100	2633

5.2 Theoretical Model Validation

We use indicator reliability and composite reliability (CR) and to validate the reliability of our model. The CR of PEOU, PU, SE, EXP, FC, COM, AQ, SI, SAF, ITU are 0.8928, 0.8221, 0.8774, 0.8698, 0.9413, 0.8801, 0.9039, 0.9115, 0.9347, respectively. The values are all higher than the recommended value of 0.70 [55].

We use the average variance extracted (AVE) to confirm the convergent validity. The AVE of PEOU, PU, SE, EXP, FC, COM, AQ, SI, SAF, ITU are 0.8100, 0.7511, 0.6930, 0.7054, 0.8870, 0.6588, 0.7836, 0.8243, 0.8330, 1.0000, respectively. All of our values was higher than 0.5, which is the standard threshold.

6 Discussion

6.1 Principal Findings

The purpose of this paper is to use the theoretical derivation process to construct a theoretical model framework of chronic disease healthcare service management. We develop an improved algorithm to enhance healthcare management. From reviewing healthcare literature and exploring the preliminary reliability and validity of our models, we find sufficient evidence to show that the computer-supported method can provide effective theoretical support and exploratory guidance for the development of chronic disease healthcare service management systems in the future.

This study highlights several challenges in previous studies. First, in the research on applying computer algorithms to the field of health services, it is necessary to pay attention to the unique nature of health services – people-centered, such as the HCP satisfaction in this model, which can directly or indirectly affect the service quality and efficiency. Second, studies on healthcare management need to consider multiple impact factors instead of focusing on a single factor. In terms of the effectiveness of use, a multi-model integration method should be adopted to balance the influence of multiple factors and to simplify the workflow.

6.2 Limitations

This study is still in the preliminary exploratory stage. Due to the small sample size of data and the absence of long-term investigation and analysis, further improvement and factor exploration are needed. At the same time, the purpose of this study is to provide theoretical support for the other design of a chronic disease health management system based on computer-supported methods. Therefore, it is necessary to explore and analyze the feasibility of system design and implementation in practice. Furthermore, as this study focused solely on HCPs, it is also essential to include patients as research subjects.

7   Conclusion

This study explores the relationship between computer-supported methods and chronic disease healthcare service management. We improve the algorithm pose-AMGRU and construct a theoretical model to verify the acceptance of our model. We have shown that the computable and multi-model fusion healthcare service model can provide academic guidance for improving the work efficiency and quality of HCP, and provide a theoretical basis for future computer technical support.

Acknowledgments. We thank all participants who provided thoughtful and constructive comments on our study. We appreciate Xiaoyi Zhang carried on the guidance of grammar problems on our paper. This research was funded by the National Key Research and Development Program of China (No. 2017YFB1401200), the General Project of Humanistic and Social Science Research of the Department of Education of Henan province. (No. 2018-ZZJH-547), and the program of China Scholarship Council (No. 201907040091).

References

1. Enthoven, A.C., Tollen, L.A.: Competition in health care: it takes systems to pursue quality and efficiency. Health Aff. (Millwood). Suppl Web 420–433 (2005)
2. Owens, G.M.: Gender differences in health care expenditures, resource utilization, and quality of care. J. Manag. Care Pharm. **14**, 2–6 (2008)
3. Parekh, A.K., Goodman, R.A., Gordon, C., Koh, H.K.: Managing multiple chronic conditions: a strategic framework for improving health outcomes and quality of life. Public Health Rep. **126**, 460–471 (2011)

4. Wennberg, J.E., Fisher, E.S., Baker, L., Sharp, S.M., Bronner, K.K.: Evaluating the efficiency of california providers in caring for patients with chronic illnesses. Health Aff. (Millwood). Suppl Web 526–543 (2005)
5. Kischka, U., Wade, D.T.: Rehabilitation after stroke. Handb. Cerebrovasc. Dis. Second Ed. Revis. Expand. 231–241 (2004)
6. Jørgensen, H.S., Nakayama, H., Raaschou, H.O., Vive-Larsen, J., Støier, M., Olsen, T.S.: Outcome and time course of recovery in stroke. Part I: outcome. The Copenhagen stroke study. Arch. Phys. Med. Rehabil. **76**, 399–405 (1995)
7. Teasell, R.W., McRae, M.P., Finestone, H.M.: Social issues in the rehabilitation of younger stroke patients. Arch. Phys. Med. Rehabil. **81**, 205–209 (2000)
8. Enthoven, A.C., Vorhaus, C.B.: A vision of quality in health care delivery. Health Aff. **16**, 44–57 (1997)
9. Feder, J., Komisar, H.L., Niefeld, M.: Long-term care in the United States: an overview. Health Aff. **19**, 40–56 (2000). https://doi.org/10.1377/hlthaff.19.3.40
10. LAST: Role of Effective Teamwork and Communication in Delivering Safe, High-Quality Care. Medicine (Baltimore), pp. 15–21 (2007)
11. Waters, T.M., Kaplan, C.M., Graetz, I., Price, M.M., Stevens, L.A., McAneny, B.L.: Patient-centered medical homes in community oncology practices: changes in spending and care quality associated with the COME HOME experience. J. Oncol. Pract. **15**, e56–e64 (2019)
12. Feo, R., Rasmussen, P., Wiechula, R., Conroy, T., Kitson, A.: Developing effective and caring nurse-patient relationships. Nurs. Stand. **31**, 54–63 (2017)
13. Lee, W., Park, J., Park, C.H.: Acceptability of tele-assistive robotic nurse for human-robot collaboration in medical environment, pp. 171–172 (2018)
14. Wang, Z.W., An, Y.: The analysis on the construction of CSCW system and group collaborative mode. Adv. Mater. Res. **756–759**, 2966–2970 (2013)
15. Stahl, G.: Theories of collaborative cognition: foundations for CSCL and CSCW together. In: Goggins, S., Jahnke, I., Wulf, V. (eds.) Computer-Supported Collaborative Learning at the Workplace. Computer-Supported Collaborative Learning Series, vol. 14, pp. 43–63. Springer, Boston (2013). https://doi.org/10.1007/978-1-4614-1740-8_3
16. Reddy, M.C., Bardram, J., Gorman, P.: CSCW research in healthcare: past, present, and future. Work. 615–616 (2010)
17. Hood, M., Wilson, R., Corsica, J., Bradley, L., Chirinos, D., Vivo, A.: What do we know about mobile applications for diabetes self-management? A review of reviews. J. Behav. Med. **39**(6), 981–994 (2016). https://doi.org/10.1007/s10865-016-9765-3
18. Kowitlawakul, Y.: The technology acceptance model: predicting nurses' intention to use telemedicine technology (eICU). CIN - Comput. Informatics Nurs. **29**, 411–418 (2011)
19. Moon, B.C., Chang, H.: Technology acceptance and adoption of innovative smartphone uses among hospital employees. Healthc. Inform. Res. **20**, 304–312 (2014)
20. Matthias, M.S., et al.: Self-management support and communication from nurse care managers compared with primary care physicians: a focus group study of patients with chronic musculoskeletal pain. Pain Manag. Nurs. **11**, 26–34 (2010)
21. Dou, K., et al.: Patients' acceptance of smartphone health technology for chronic disease management: a theoretical model and empirical test. JMIR mHealth uHealth. **5**, e177 (2017)
22. Schüldt, C., Laptev, I., Caputo, B.: Recognizing human actions: a local SVM approach. Proc. - Int. Conf. Pattern Recognit. **3**, 32–36 (2004)
23. Cao, Z., Simon, T., Wei, S.E., Sheikh, Y.: Realtime multi-person 2D pose estimation using part affinity fields. In: Proceedings - 30th IEEE Conference on Computer Vision and Pattern Recognition, CVPR 2017, vol. 2017-Janua, pp. 1302–1310 (2017)
24. Rowe, N.C., Chan, A.L.: On the properties of neural machine translation: encoder-decoder approaches. In: Proceedings of 2011 International Conference on Image Processing, Computer Vision, Pattern Recognition, IPCV 2011, vol. 1, pp. 317–322 (2011)

25. Liu, W., et al.: SSD: Single shot multibox detector. In: Leibe, B., Matas, J., Sebe, N., Welling, M. (eds.) ECCV 2016. LNCS, vol. 9905, pp. 21–37. Springer, Cham (2016). https://doi.org/10.1007/978-3-319-46448-0_2

26. Huang, G., Liu, Z., Van Der Maaten, L., Weinberger, K.Q.: Densely connected convolutional networks. In: Proceedings - 30th IEEE Conference on Computer Vision and Pattern Recognition, CVPR 2017. 2017-Janua, 2261–2269 (2017).

27. Davis, F.D., Bagozzi, R.P., Warshaw, P.R.: User acceptance of computer technology: a comparison of two theoretical models. Manage. Sci. 35, 982–1003 (1989). https://www.jstor.org/stable/2632151

28. Yarbrough, A.K., Smith, T.B.: A new take on TAM. Med. Care Res. Rev. 64, 650–672 (2007)

29. Ooi, K.B., Tan, G.W.H.: Mobile technology acceptance model: an investigation using mobile users to explore smartphone credit card. Expert Syst. Appl. 59, 33–46 (2016)

30. Müller, J.M.: Comparing technology acceptance for autonomous vehicles, battery electric vehicles, and car sharing-a study across Europe, China, and North America. Sustain. 11, 4333 (2019)

31. Hu, P.J., Chau, P.Y.K., Sheng, O.R.L., Tam, K.Y.: Examining acceptance model using physician of acceptance telemedicine technology. J. Manag. Inf. Syst. 16, 91–112 (2012)

32. Park, Y., Chen, J.V.: Acceptance and adoption of the innovative use of smartphone. Ind. Manag. Data Syst. 107, 1349–1365 (2007)

33. Lee, Y., Kozar, K.A., Larsen, K.R.T.: The technology acceptance model: past, present, and future. Commun. Assoc. Inf. Syst. 12, 50 (2003)

34. Chau, P.Y.K., Hu, P.J.-H.: Information technology acceptance by individual professionals: a model comparison approach. Decis. Sci. 32, 699–719 (2007)

35. Godin, G., Bélanger-Gravel, A., Eccles, M., Grimshaw, J.: Healthcare professionals' intentions and behaviours: a systematic review of studies based on social cognitive theories. Implement. Sci. 3, 1–2 (2008)

36. Venkatesh, M., Davis, D.: User acceptance of information technology: toward a unified view. MIS Q. 27, 425 (2003). https://doi.org/10.2307/30036540

37. Jewer, J.: Patients' intention to use online postings of ED wait times: a modified UTAUT model. Int. J. Med. Inform. 112, 34–39 (2018)

38. Krishnan, G., Mintz, J., Foreman, A., Hodge, J.C., Krishnan, S.: The acceptance and adoption of transoral robotic surgery in Australia and New Zealand. J. Robot. Surg. 13(2), 301–307 (2018). https://doi.org/10.1007/s11701-018-0856-8

39. Cimperman, M., Makovec Brenčič, M., Trkman, P.: Analyzing older users' home telehealth services acceptance behavior-applying an Extended UTAUT model. Int. J. Med. Inform. 90, 22–31 (2016)

40. Ramirez, M., Wu, S., Ryan, G., Towfighi, A., Vickrey, B.G.: Using beta-version mhealth technology for team-based care management to support stroke prevention: an assessment of utility and challenges. JMIR Res. Protoc. 6, e94 (2017)

41. Carayon, P., et al.: Work system design for patient safety: the SEIPS model. Qual. Saf. Heal. Care. 15, 50–58 (2006)

42. Dul, J., et al.: A strategy for human factors/ergonomics: developing the discipline and profession. Ergonomics 55, 377–395 (2012)

43. Rublee, D.A.: The quality of care: how can it be assessed? JAMA J. Am. Med. Assoc. 261, 1151 (1989)

44. Wooldridge, A.R., Carayon, P., Hundt, A.S., Hoonakker, P.L.T.: SEIPS-based process modeling in primary care. Appl. Ergon. 60, 240–254 (2017)

45. Zhou, M., Zhao, L., Kong, N., Campy, K.S., Qu, S., Wang, S.: Factors influencing behavior intentions to telehealth by Chinese elderly: an extended TAM model. Int. J. Med. Inform. 126, 118–127 (2019)

46. Yu, P., Li, H., Gagnon, M.P.: Health IT acceptance factors in long-term care facilities: a cross-sectional survey. Int. J. Med. Inform. **78**, 219–229 (2009)

47. Carayon, P., et al.: Human factors systems approach to healthcare quality and patient safety. Appl. Ergon. **45**, 14–25 (2014)

48. Razmak, J., Bélanger, C.H., Farhan, W.: Development of a techno-humanist model for e-health adoption of innovative technology. Int. J. Med. Inform. **120**, 62–76 (2018)

49. Todd, P.A., Taylor, S.: Understanding information technology usage: a test of competing models. Inf. Syst. Res. **6**, 144–176 (1995). https://www.jstor.org/stable/23011007. Understanding Models Usage

50. Schoot, T., Zuyd, H.: Client-centred care balancing between perspectives (2015)

51. Ji, S., Xu, W., Yang, M., Yu, K.: 3D Convolutional neural networks for human action recognition. IEEE Trans. Pattern Anal. Mach. Intell. **35**, 221–231 (2013)

52. Megrhi, S., Jmal, M., Souidene, W., Beghdadi, A.: Spatio-temporal action localization and detection for human action recognition in big dataset. J. Vis. Commun. Image Represent. **41**, 375–390 (2016)

53. Yuxi, M., Li, T., Dong, X., et al.: Action recognition for intelligent monitoring. J. Image Graph. **24**(02), 128–136 (2019)

54. Sargano, A.B., Wang, X., Angelov, P., Habib, Z.: Human action recognition using transfer learning with deep representations. In: Proceedings of International Joint Conference on Neural Networks, vol. 2017-May, pp. 463–469 (2017)

55. Guo, X., Sun, Y., Wang, N., Peng, Z., Yan, Z.: The dark side of elderly acceptance of preventive mobile health services in China. Electron. Mark. **23**, 49–61 (2013)

Multiple Interpretations Improve Deep Learning Transparency for Prostate Lesion Detection

Mehmet A. Gulum[(✉)], Christopher M. Trombley, and Mehmed Kantardzic

University of Louisville, Louisville, KY 40203, USA
{mehmetakif.gulum,christopher.trombley,mehmed.kantardzic}@louisville.edu

Abstract. Detecting suspicious lesions in MRI imaging is a critical task in preventing deaths from cancer. Deep learning systems have produced remarkable accuracy for the task of detecting lesions in MRI images. Although these systems show remarkable performance, they often ignore an indispensable component which is interpretability. Interpretability is essential for many deep learning applications in medicine because of ethical, monetary, and legal factors. Interpretation also builds a necessary degree of trust and transparency between the doctor, patient, and system. This work proposes a framework for the interpretation of medical deep learning systems. The proposed approach is based on the idea that it is advantageous to use different interpretation techniques to show multiple views of reasoning behind the classification. This work demonstrates deep learning interpretations for various patient data modalities using the proposed Multiple Views of Interpretation for Deep Learning framework.

Keywords: XAI · Deep learning · Prostate cancer · Interpretation · Visualization

1 Introduction

There have been recent advances using deep learning techniques, such as convolutional neural networks [1], to detect prostate cancer from MRI images with impressive performance. [2] showed deep learning can detect prostate cancer with accuracy suitable to be integrated into a clinical environment. These results demonstrate the great potential for using deep learning to aid medical practitioners. However, these advances often ignore an indispensable component of such systems which is interpretability.

Although deep learning models can produce accurate cancer classification, they are often treated as black-box models that lack interpretability and transparency of their inner working [3]. The models provide an accurate classification but do not demonstrate how they arrived at the decision. If such systems are to be implemented into medical settings, integrating interpretability is an essential, often overlooked component. Interpretability is needed for various reasons. First,

V. Gadepally et al. (Eds.): Poly 2020/DMAH 2020, LNCS 12633, pp. 120–137, 2021.
https://doi.org/10.1007/978-3-030-71055-2_11

there are legal and ethical requirements along with laws and regulations that are required for deep learning cancer detection systems to be implemented in a clinical setting. An example of a regulation is the European Union's General Data Protection Regulation (GDPR) requiring organizations that use patient data for classifications and recommendations to provide on-demand explanations [4]. The inability to provide such explanations on demand may result in large penalties for the organizations involved. Thus, there are monetary incentives associated with interpretable deep learning models. Beyond ethical and legal issues, clinicians and patients need to be able to trust the classifications provided by these systems. Interpretation attempts to show the reasoning behind the model's classification thus building a degree of trust between the system, clinician, and patient. Theoretically, this will reduce the number of misdiagnosed cases that would be a possible consequence of non-interpretable systems. Finally, interpretable deep learning systems will provide the clinician with practical features as a second-order effect. Examples of these practical features are the ability to provide segmentation of a medical region of interest (ROI) [5] and the localization of lesions [6].

Interpretation methods can be categorized as post-hoc and ad-hoc. Post-hoc refers to interpretation after the classification is made whereas ad-hoc refers to engineering interpretation into the deep learning system. This work will largely focus on post-hoc approaches. There are various types of interpretation techniques that highlight different aspects of classification for the same sample. Some of them highlight the localization of a lesion and others highlight the size or area of a tumor or cyst. This paper will shed light on the importance of interpretation for medical deep learning systems, the current state of interpretation for deep learning within a medical context, and will propose a viable approach for medical deep learning interpretation moving forward. The main contributions of this paper are (1) showing that the integration of multiple interpretation techniques produce a new quality and delivers greater insight into the model's classification opposed to using a single method (2) establish an evaluation methodology for measuring visual interpretation performance (3) demonstrate that Grad-CAM can precisely localize prostate lesions in T2W and ADC MRI images.

2   Related Work

2.1   Classification of Prostate Cancer and Lesions

Deep learning has been widely applied to the classification of medical conditions ranging from diabetes to cancer. There are attempts to use deep learning techniques to detect cancer, some of which produce remarkable performance. [7] demonstrate a transfer learning approach to detect prostate lesions using MRI images. The study implements the InceptionV3 and VGG16 models which were both initialized with imagenet weights for the task of detecting prostate lesions using the PROSTATEx dataset [8] Transfer learning is used in many cancer detection systems because of the advantage from initializing a network with pre-trained parameters. Ensemble learning techniques were implemented

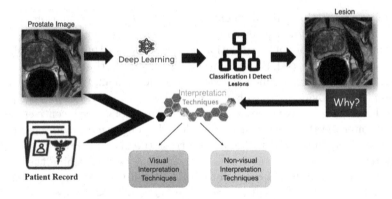

Fig. 1. Global overview of work

to improve the area under the curve. Their results range from 0.82 to 0.91 for AUROC. [9] implement a three-dimensional convolutional neural network for the task of classifying clinically significant lesions. The study reports an AUROC of 0.80 and argues that with the 3D network, spatial information is captured thus producing a model with greater insight into 3D medical volumes. [10] proposed a method called XmasNet. Their work performs novel data augmentation using three-dimensional rotation and slicing, in order to incorporate the three-dimensional information of the lesion volume. The study reports an AUC of 0.92. [11] propose a fully automated approach to prostate lesion detection using MRI images reporting an AUROC of 0.84 (Fig. 1).

2.2   Post-Hoc Interpretation for Deep Learning in Medicine

Post-hoc interpretation attempts to provide reasoning after the classification is made as opposed to engineering interpretation into the deep learning system. These post-hoc visual interpretation techniques generally either use perturbation forward propagation, backward propagation, or gradient-based visual explanation methods. Perturbation forward propagation make perturbations to individual inputs or neurons and observe the impact on later neurons in the network. Backward propagation is the opposite. Instead of propagating forward through the network, a signal is propagated from the output neuron(s) back to the input neurons. Gradient-based methods propose taking the gradient of an output variable with respect to input variables to calculate which input variables change the outcome the most. [12] examine post-hoc interpretation for the classification of melanoma in histology slides. Their work trains a ResNet50 and VGG19 using transfer learning to for the classification of melanoma. Class activation map (CAM) [13] is used to provide a post-hoc visual explanation. [14] design a neural network to perform analysis on frames collected from an endoscopic examination taken from a video stream. This uses Grad-CAM and guided Grad-CAM as gradient-based post-hoc interpretation techniques to explain findings in these frames. [15] show post-hoc interpretability using LIME for clinical data.

[16] show post-hoc interpretability using LIME for acute kidney injury in cardiac surgery patients. Their work uses LIME to attempt to explain the onset of the condition. [17] created a deep learning system to classification hypertension and then used LIME to explain these classifications. [18] developed a model to classifications diabetic retinopathy progression in individual patients. They then use SHAP to give insight into the model's decisions.

Emerging literature has highlighted the ability to use these interpretation techniques to localize lesions or other medical regions of interest. [19] use a modified version of Grad-CAM, coined pyramid gradient-based class activation mapping (PG-CAM), to localize meningioma. They report a 23% increase from vanilla Grad-CAM for the localization of brain tumors. [20] propose high-resolution CAM (HR-CAM) which aggregates feature maps together. They localize ependymomas using this technique. [21] use saliency maps to segment lesions in dermoscopy images with a DICE coefficient of 0.858. To the best of our knowledge, there are not any studies that localize prostate lesions from MRI images using interpretation techniques.

3 Multiple Views of Interpretations for Deep Learning

Interpretation methods (i.e. Grad-Cam, LIME, SHAP, saliency maps) show a certain aspect of the reasoning behind a deep learning model's classification. Each interpretation method by itself provides some method-specific information. For example, grad-CAM highlights an area of interest whereas LIME highlights which clinical features contribute the most to the classification. Saliency maps show the important structure and clusters of important individual pixels. These individual interpretation techniques can be considered individual parts of a larger system. By combining these methods together, a higher quality interpretation is produced.

Using a multiple interpretation approach is advantageous for multiple reasons. First, the clinician and patient will receive more insight into the model's classification using a combination of methods as opposed to a sole method. Second, you gain a higher degree of confidence using an approach which includes multiple interpretations. If the different methods are in unison, then the consistency delivers a degree of confidence in the interpretation and classification. Third, it is possible for interpretations can be fooled and produce misleading interpretations. Using multiple interpretations approach, theoretically, you can uncover issues with the classification because of a lack of consistency between interpretations. Lastly, you can provide an interpretation for many different data modalities (i.e. images, genetic information, clinical information, patient history) which is essential for healthcare applications.

This work presents a framework for interpretation, Multiple Views of Interpretation for Deep Learning, for prostate lesion detection and interpretation (Fig. 2). Combining multiple interpretation methods will increase transparency and give a multifaceted view into how the model arrives at the classification thus providing a holistic interpretation.

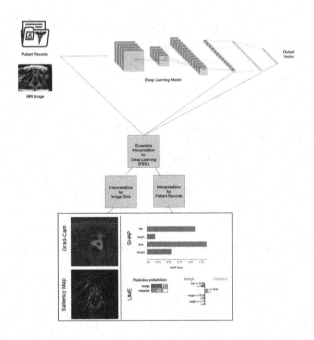

Fig. 2. MVIDL architecture for prostate lesion detection.

3.1 Deep Learning Model

In this paper, convolution neural network (CNN) based on the VGG16 [22] was implemented with some improvements. InceptionV3, VGG16, VGG19, ResNet50, MobileNet, and WideResNet were all implemented and compared to select the optimal model. Each network's hyper-parameters were tuned using grid search. After tuning, the VGG16 produced the highest performance. Two extra convolutional layers followed by max-pooling were concatenated to the end for increased classification performance. The network is designed to classify individual slices with a lesion from slices without a lesion. Clinical features are concatenated to the fully connected layers to incorporate patient records. The hyper-parameters after tuning were: 18 layers, a weight decay of 0.00001, a learning rate of 0.001, the ADAM optimizer, the binary cross-entropy loss function, a softmax activation as the final layer, and the network was initialized with imagenet weights (Fig. 3).

VGG-Net Architecture

Fig. 3. VGG16 architecture

3.2 Interpretation Techniques for Prostate Lesion Detection

Combining interpretation techniques produces a more transparent system by giving a holistic view of the model's decision. Each technique gives unique insight, therefore, to get a comprehensive interpretation, each technique is needed as individual parts of a larger system. For this work, these techniques are split into two categories: image data interpretation and clinical data interpretation. Image data refers to MRI imaging in this work and clinical data refers to patient characteristics including weight, age, height, and body mass index (BMI). The techniques for image interpretation are focused solely on providing explanations for image data thus do not take into account clinical information. The second category is clinical data interpretation which takes into account clinical information but does not provide an explanation for image data. This work shows Grad-CAM [23] and saliency maps [24,25] as the techniques for image data. SHAP [26] and LIME [27] as the techniques for clinical data. Each technique will be explained in detail below.

Gradient-weighted Class Activation Mapping (Grad-CAM) uses the gradients flowing into a convolutional layer to produce a map that highlights important regions in the image for the classification of a class.

$$\frac{\partial y^c}{\partial A^k} \tag{1}$$

Grad-CAM calculates the gradients of an individual class score, c, with respect to feature map, A.

$$\alpha_k^c = \frac{1}{Z} \sum_i \sum_j \frac{\partial y^c}{\partial A_{ij}^k} \tag{2}$$

Neuron importance is calculated by globally average pooling the gradients.

$$L_{Grad-CAM}^c = ReLU(\sum_k \alpha_k^c A^k) \tag{3}$$

A weighted combination of forward activation maps is followed by a ReLU to obtain the final activation map. The advantage of Grad-CAM is it takes into account feature maps thus showing how well your model learns quality features. This is important when training a model for clinical diagnosis because you can examine if your model is learning interpretable features using Grad-CAM. A disadvantage is the heatmaps can be unclear.

Saliency maps show the individual pixels that contribute the most to the class score. This is useful for showing the structures and clusters of pixels that contribute the most to the class score. Mathematically, saliency maps calculate the partial derivative of the class score with respect to an individual pixel at a specific pixel.

$$w = \frac{\partial S_c}{\partial I} \tag{4}$$

It repeats this process for each pixel in the image and assigns each pixel a numeric value. This value represents the contribution to the class score for each pixel. You will notice in the results, the highlighting of the structure as well as pixel clustering around 'important' areas in the image. The advantage of saliency maps is they highlight individual pixels that are important thus providing a precise interpretation. The disadvantage is individual pixels may not matter as much as clusters of pixels (i.e. feature maps).

Local Interpretable Model-agnostic Explanations (LIME) creates an interpretable model locally around a classification. It produces a bar graph showing the contribution of each feature from the patient records. Each bar shows the direction and magnitude of contribution.

The explanation produced by LIME is obtained by the following:

$$\xi(x) = argminL(f, g, \pi_\infty) + \Omega(g) \tag{5}$$

Where f is a machine learning model, g is the explanation defined as a model, $pix(z)$ is used as a proximity measure between an instance z to x, so as to define locality around x, and $omega(g)$ is a measure of complexity. The advantages of LIME are it produces a clear and concise graph that is easily interpretable. The disadvantage is the method does not support multi-modal input.

SHAP (SHapley Additive exPlanations is a unified, model-agnostic approach to interpretation based on game theoretically optimal Shapley Values. The way SHAP calculates feature importance is as follows.

$$g(z) = \phi_0 + \sum_{j=1}^{M} \phi_j z_j \tag{6}$$

Where g is the explanation model, z is the coalition vector, M is the maximum coalition size, and phi is the feature attribution for feature j. For global importance, we average the absolute Shapley values per feature across the data as shown below. The advantage of SHAP is that it provides a model-agnostic, personalized global and local interpretation.

$$I_j = \sum_{i=1}^{n} \left| \phi_j^{(i)} \right| \tag{7}$$

4 Dataset

The dataset used for this work was the PROSTATEx dataset [28] from the SPIE-AAPM-NCI PROSTATEx challenge. The PROSTATEx dataset consists of 330 lesions from 204 patients. The dataset provides DICOM coordinates for the centroid of the prostate lesions. The lesions in the dataset were labeled as clinically significant or not clinically significant depending on their pi-rads score. The dataset provides T2W transaxial, T2W sagittal, T2W coronal, ADC, and BVAL image modalities. This study includes T2W and ADC images. Six patients were excluded from this work due to poor image quality. This results in 199 patients and 322 lesions. The test set for both classification and interpretation includes 103 images of lesions and 103 images without lesions.

The data preprocessing steps are as follows: The T2W images are downsized from 350×350 to 224×224 pixels. The ADC images are downsized from 120×80 to 50×50 pixels. All images are then converted to RGB images. The pixel values are normalized using z-scoring. Then data augmentation is carried out using shearing, rotation, and translating data augmentation techniques. The lesion centroid coordinates are then converted to the resized coordinate frame using the following formulas:

$$x_{new} = x_{old} \times \frac{x_{current}}{width_{current}} \qquad y_{new} = y_{old} \times \frac{y_{current}}{height_{current}} \qquad (8)$$

5 Experimental Results

This section is split into four sub-sections: classification results, image interpretation results, clinical data interpretation results, and Multiple Views for Interpretation for Deep Learning results. In Sect. 5.1, lesion classification results are introduced after tuning the parameters of methodology with ProstateX data set. The image interpretation shows the different visualizations used to gain insight into the classifications. This part also demonstrates the precision of the localization of prostate lesions using image interpretation technique (Grad-Cam) in Sect. 5.2. In Sect. 5.3, clinical data interpretation results demonstrate local and global clinical data interpretation with LIME and SHAP interpretation techniques. Lastly, the advantages of using multiple interpretation techniques are demonstrated in Sect. 5.4.

5.1 Classification Result

The classification results are shown in Table 1. These results demonstrate that engineering interpretation into deep learning can still produce models with classification performance. As can we see on Table 1, our work (VGG Net) has almost similar result when we compare with previous works for accuracy. Although XmasNet has better values, our results are close enough to go second step which are interpretation techniques. These results also show that true positives and false positives are captured using this approach. False negatives and false positives are likely to be correctly filtered out and classified correctly. It provides a credible model with results comparable to models in relevant literature to demonstrate the interpretation techniques.

Table 1. Lesion classification results

Method	AUC	Sensitivity	Specificity
XmasNet [10]	0.92	0.89	0.89
DCNN [11]	0.84	0.69	0.83
3DCNN [9]	0.85	–	–
VGG Net	0.84	0.81	0.86

5.2   Evaluation of Grad-Cam Precision Results

Grad-CAM highlights the area of the image that contributes most to the classification. An additional finding is this highlights the lesion centroid in MRI images with high precision. To measure interpretation quality for Grad-CAM, the assumption is made that a credible interpretation would highlight the lesion centroid as the most important area. That is, if the slice contained a lesion in the ground truth. We propose the following performance measures for interpretation for image data: the distance between centroids, false positives, false positive, correctly localized, and incorrectly localized.

Incorrectly localized interpretations are measured as the number of samples that produce a heatmap that does not accurately highlight lesion, see Fig. 6. The heatmap is considered correctly localized if the centroid of the lesions falls within the radius of the heatmap. False Positives are measured as the number of cases that produce a heatmap given a slice without a lesion, see Fig. 5. False Negatives are measured as the number of cases that do not show a heatmap given an input slice that contains a lesion, see Fig. 4.

The interpretation is considered correctly localized if the coordinates of the lesion centroid are located within the radius of the heat map. If this does not hold true then it is considered incorrectly localized. The threshold is the radius of the heat map which varies from 5 pixels to 16 pixels. Table 2 shows the errors for Grad-CAM visualizations in terms of false positives, false negatives, and incorrect localization. Examples of false negatives, false positives, and incorrect localization are shown in Figs. 4, 5, and 6 respectively.

Distance is calculated between the centroid of the lesion and the geometric center of the heatmap using the distance formula shown below:

$$d = \sqrt{(x_2 - x_1)^2 + (y_2 - y_1)^2} \tag{9}$$

103 images were analyzed and the precision of lesion localization was calculated using Grad-CAM. T2W images were 224×224 pixels and ADC images were

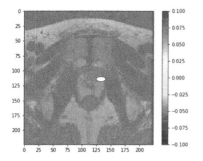
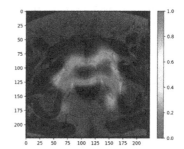

Fig. 4. False Negative example.   Fig. 5. False Positive example.

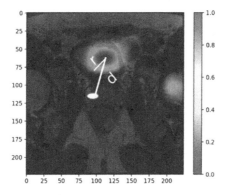

Fig. 6. Incorrectly localized example.

50×50 pixels. Figure 7 and Fig. 8 show the precision of Grad-Cam for T2W and ADC images respectively. The interpretation results for Grad-CAM, measured as a localization task, for image data interpretation are a mean distance of 6.93 for T2W images and a mean distance of 16.3 for ADC images. These results do not only show that the interpretation is clear for clinicians, but also that this method can precisely localize lesion location.

Grad-CAM is useful when you want the interpretation to highlight high-level (i.e. human interpretable) features. Examples of this would be tumor

Table 2. Grad Cam results

Error	T2	ADC
False negatives	3	5
False positives	17	21
Incorrect localization	12	17
Correct localization	174	163

classification and neuroimaging studies. This method is also useful because of its ability to localize a region of interest such as a prostate lesion. This method does not work as well if you want to highlight the boundaries precisely.

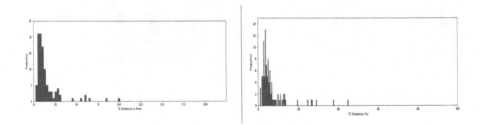

Fig. 7. Histogram for T2W lesions localization using Grad-Cam.(pixel-wise and percentage)

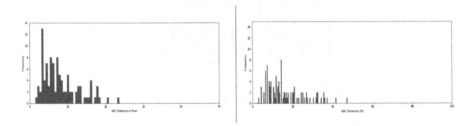

Fig. 8. Histogram for ADC lesions localization using Grad-Cam.(pixel-wise and percentage)

Figure 7 and Fig. 8 show how T2W and ADC images distribute precision of Grad-Cam in percentage and pixel-wise. Based on these figures, the average sample is more likely to be localized accurately in T2W images opposed to ADC images. The inaccuracy in ADC images can be contributed to significantly lower image resolution (Table 3).

Table 3. Precision of Grad-Cam

Statistic	T2	ADC
Mean distance	6.93	16.3
Standard deviation	7.4	9.5

Saliency maps show the importance of the structure within the image. They also show which pixels are important for the classification. Thus with saliency maps, you can examine where the clusters of pixels are. These clusters are the areas the model considers most important. This is how saliency maps differ from Grad-CAM. Grad-CAM focuses on feature maps whereas saliency maps focus on individual pixels. If the cluster of pixels is in the same region as the Grad-CAM heatmap that shows consistency between methods and instills confidence in the classification. This provides a sense of trust for the interpretation that the area of the image contributes most to the classification.

Saliency maps show the structure of the image well. Also, they are useful if you want to outline an object within the image. For example, if you want to extract tumor shape or orientation. Saliency maps work well if you want to visualize which pixels contribute the most to the classification. You can gain insight into the important regions by examining where the clusters of pixels are. This does not work well if you want to produce a clear, concise interpretation because this method is often unclear.

5.3   Interpretation Results for Clinical Data

In previous sections, we mentioned visual results with interpretation techniques. In this part, non-visual interpretation techniques which are LIME and SHAP were tested for patient record. For LIME, the model shows BMI and age are most important for the personalized interpretation. The SHAP values representing global feature importance are consistent with this showing that BMI and age the most important features globally. This shows consistency between the local and global methods thus providing a sense of trust that the interpretation is accurate. These results show which clinical features contribute most to the classification.

LIME is useful if you want individualized interpretation. An advantage of LIME is the graphs it provides. They show not only magnitude, but also direction of feature importance. They also are color coded and organized in a clear and concise manner. If you want to examine global feature importance it is better to not use LIME (Figs. 9 and 10).

Fig. 9. LIME results for a patient with a benign lesion.

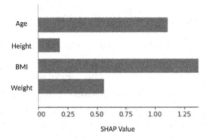

Fig. 10. LIME results for a patient with a malignant lesion.

SHAP is useful if you want to examine global feature importance along with local feature importance. SHAP is also useful because of it's model-agnostic feature. It can show reasoning behind the classification regardless of what type of model you are using. SHAP also provides straightforward interpretation using SHAP value. Future work can include more variables for these interpretation methods such as patient history, genetic information, additional patient characteristics, and anything else the clinician deems appropriate (Fig. 11).

Fig. 11. Global feature importance depicted as absolute SHAP value.

5.4 Multiple Views for Interpretation for Deep Learning Results

In this section, two cases are shown for end-to-end personalized interpretations using the Multiple Views for Interpretation for Deep Learning Results framework. This shows what regions of the images contribute to the classification, how the structure influences classification, what cluster of pixels are important, what global features matter the most across the entire cohort, and what clinical information contributes to the system's decision. This shows that using multiple interpretation techniques provides more complete reasoning behind the prediction as opposed to using a single technique. Using Multiple Views for Interpretation for Deep Learning, we show: (1) what areas of the image contribute most the prediction (2) the importance of the structure of the prostate and surrounding area (3) what individual pixels, and clusters of pixels, contribute the most (4) what features from the patient's medical records contribute the most (5) localization of the lesion and (6) what medical record features contribute the most globally across the cohort. This provides a more complete explanation than an individual technique. This combines the strength of each individual method to create a greater whole. There is consistency between the different image interpretation methods. This shows that the area where the heatmap is, and the pixels are clustered, is the most important region of the image for the classification. Since both interpretation techniques are consistent, there is a degree of trust in the reasoning the model gives for the classification. For clinical data interpretation, the local interpretations are consistent with global interpretations. This provides confidence the neural network is using sound reasoning to make life-critical classifications. In Figs. 12, 13, 14 and 15, each example is tested using image and patient records from the PROSTATEx dataset. Three patient were selected and tested with our framework. The first patient has a age, weight, BMI, and height of 58 years, 70 kg, 23 kg/m2 and 176 cm. The second patient has a age, weight, BMI, and height of 75 years, 80 kg, 28 kg/m^2, and 200 cm.

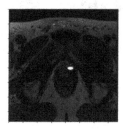

Feature Values	
BMI	23.00
Age	58.00
Height	1.76
Weight	70.00

Original T2W MRI Image   Non-Visual Features

Fig. 12. Patient I Inputs. The white dot in image shows the lesion centroid location.

Based on Fig. 12 and 13, the heatmap shows the area of the image that contributes the most to the class score. This heatmap highlights the lesion centroid. Saliency map shows the individual pixels that contribute the most to the class score. You can see a cluster of pixels at the lesion centroid. You can also see how it highlights the edges, shapes, and structure of the prostate area. SHAP shows global feature importance.

LIME shows the clinical information features that contribute the most to the class score. For this patient, his bmi and age contribute the most. The bmi is low so even though the network says the age contributes towards malignant, the bmi is low enough to give confidence he is healthy.

Based on Fig. 14 and 15, the heatmap shows the area of the image that contributes the most to the class score. This heatmap highlights the lesion centroid. Saliency map shows the individual pixels that contribute the most to the class score. You can see a cluster of pixels at the lesion centroid. You can also see how it highlights the edges, shapes, and structure of the prostate area. SHAP shows global feature importance. LIME shows the clinical information features that contribute the most to the class score. For this patient, his bmi and age contribute the most. The bmi and age are both high so the network considers this patient malignant overall.

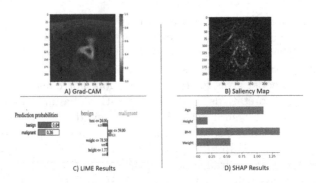

Fig. 13. Multiple views for Interpretation for Deep Learning Results for Patient I.

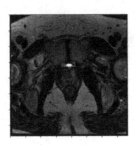

Feature Values	
BMI	28.00
Age	75.00
Height	2.00
Weight	80.00

Original T2W MRI Image   Non-Visual Features

Fig. 14. Patient II Inputs. The white dot in image shows the lesion centroid location.

6   Discussion and Future Work

This work proposes Multiple Views for Interpretation for Deep Learning which is an interpretation framework for deep learning medical systems. We utilize a deep convolution neural network for the task of image classification. This is used to demonstrates classification performance greater than, or on par with, similar prostate cancer classification models in the literature. Clinical information is concatenated to the fully connected layers of the CNN. Thus, we use image data (i.e. MRI images) and clinical data (i.e. patient information). This model is then used to show that multiple interpretation techniques gives greater insight compared to a single interpretation method. The network and interpretation methods are trained and tested on the PROSTATEx dataset because of the number of images and availability of patient information and lesion centroid location. The framework is extendable to other data modalities such as clinical notes or genetic data. The four methods included are Grad-CAM, saliency maps, SHAP, and LIME. Using Grad-CAM we show, not only that one can gain insight into the model's classifications, but also precisely localize lesion location using this technique. Saliency maps show the importance of the structure and individual pixels that contribute most to the class score. The results from Grad-CAM are consistent with clusters of pixels from saliency maps showing agreement between

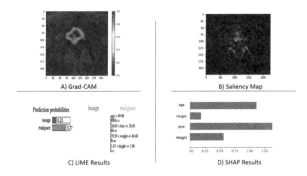

Fig. 15. Multiple Views for Interpretation for Deep Learning Results for Patient II.

image interpretation methods. For clinical data interpretation, LIME is used to show personalized classification using patient information such as weight, age, height, and BMI. SHAP is used to show global feature importance which can be used when examining LIME's personalized interpretation to determine if the classification is reasonable. These techniques are then integrated to provide a multifaceted approach to deep learning interpretation within a medical context. An interesting additional finding is using Grad-CAM we were able to accurately localize lesions in slices that contained a lesion. This work shows and analyzes different approaches to handle interpretability, one of the problems that come along with computer-aided diagnosis systems. Using multiple interpretations, the number of incorrect diagnoses influenced by deep learning systems can be reduced. The legal issues that come along with these systems will be mitigated. The systems will be more successful and ethical in practice. This makes using these systems more ethical because the reasoning will enable these systems to work alongside clinicians as opposed to carrying out their tasks. Lastly, they will ensure a degree of trust and credibility by showing the reasoning behind the model's decision. Most importantly, more work needs to be done to validate interpretation approaches in clinical settings and testing the generalizability of interpretation methods. Future work should also include working with clinicians to tailor interpretation methods to suit their specific needs. This future work also should study the integrity of such systems. If deep learning cancer detection systems are going to be implemented into clinical settings, it is of utmost importance that we trust the classifications. An approach to this is to implement multiple integrated post-hoc interpretability into these systems to provide a holistic interpretation of the model's decision. The clinical validation of this hypothesis is an important future direction.

References

1. LeCun, Y., Haffner, P., Bottou, L., Bengio, Y.: Object recognition with gradient-based learning. Shape, Contour and Grouping in Computer Vision. LNCS, vol. 1681, pp. 319–345. Springer, Heidelberg (1999). https://doi.org/10.1007/3-540-46805-6_19
2. Antonelli, M., Johnston, E.W., Dikaios, N., et al.: Machine learning classifiers can classification Gleason pattern 4 prostate cancer with greater accuracy than experienced radiologists. Eur. Radiol. **29**, 4754–4764 (2019). https://doi.org/10.1007/s00330-019-06244-2
3. Lipton, Z.C.: The mythos of model interpretability. CoRR, abs/1606.03490 (2016). http://arxiv.org/abs/1606.03490
4. Hoofnagle, C.J., van der Sloot, B., Borgesius, F.Z.: The European Union general data protection regulation: what it is and what it means. Inf. Commun. Technol. Law **28**(1), 65–98 (2019). https://doi.org/10.1080/13600834.2019.1573501
5. Canalini, L., Pollastri, F., Bolelli, F., Cancilla, M., Allegretti, S., Grana, C.: Skin lesion segmentation ensemble with diverse training strategies. In: Vento, M., Percannella, G. (eds.) CAIP 2019. LNCS, vol. 11678, pp. 89–101. Springer, Cham (2019). https://doi.org/10.1007/978-3-030-29888-3_8
6. Yoon, H.J., et al.: A lesion-based convolutional neural network improves endoscopic detection and depth classification of early gastric cancer. J. Clin. Med. **8**(9), 1310 (2019). https://doi.org/10.3390/jcm8091310
7. Chen, Q., Hu, S., Long, P., Lu, F., Shi, Y., Li, Y.: A transfer learning approach for malignant prostate lesion detection on multiparametric MRI. https://doi.org/10.1177/1533033819858363
8. Litjens, G., Debats, O., Barentsz, J., Karssemeijer, N., Huisman, H.: Computer-aided detection of prostate cancer in MRI. IEEE Trans. Med. Imaging **33**, 1083–1092 (2014). https://doi.org/10.1109/TMI.2014.2303821
9. Mehrtash, A., et al.: Classification of clinical significance of MRI prostate findings using 3D convolutional neural networks. In: Proceedings of SPIE-the International Society for Optical Engineering, vol. 10134, p. 101342A (2017). https://doi.org/10.1117/12.2277123
10. Liu, S., Zheng, H., Feng, Y., Li, W.: Prostate cancer diagnosis using deep learning with 3D multiparametric MRI. In: Armato, S.G., Petrick, N.A. (eds.) Medical Imaging (2017)
11. Wang, X., Yang, W., Weinreb, J., et al.: Searching for prostate cancer by fully automated magnetic resonance imaging classification: deep learning versus non-deep learning. Sci. Rep. **7**, 15415 (2017)
12. Xie, P., Zuo, K., Zhang, Y., Li, F., Yin, M., Lu, K.: Interpretable classification from skin cancer histology slides using deep learning: a retrospective multicenter study (2019)
13. Zhou, B., Khosla, A., Lapedriza, A., Oliva, A., Torralba, A.: Learning DeepFeatures for discriminative localization. In: CVPR (2016)
14. Hicks, S., et al.: Dissecting deep neural networks for better medical image classification and classification understanding. In: 2018 IEEE 31st International Symposium on Computer-Based Medical Systems (CBMS), Karlstad, pp. 363–368 (2018)
15. Zhang, Z., et al.: Opening the black box of neural networks: methods for interpreting neural network models in clinical applications. Ann. Transl. Med. **6**(11), 216 (2018). https://doi.org/10.21037/atm.2018.05.32

16. Da Cruz, H.F., et al.: Classification of acute kidney injury in cardiac surgery patients: interpretation using local interpretable model-agnostic explanations. In: HEALTHINF (2019)
17. Elshawi, R., Al-Mallah, M.H., Sakr, S.: On the interpretability of machine learning-based model for predicting hypertension. BMC Med. Inform. Decis. Mak. **19**, 146 (2019). https://doi.org/10.1186/s12911-019-0874-0
18. Arcadu, F., Benmansour, F., Maunz, A., et al.: Deep learning algorithm predicts diabetic retinopathy progression in individual patients. NPJ Digit. Med. **2**, 92 (2019)
19. Lee, S., et al.: Robust tumor localization with pyramid Grad-CAM. arXiv abs/1805.11393 (2018)
20. Shinde, S., Chougule, T., Saini, J., Ingalhalikar, M.: HR-CAM: precise localization of pathology using multi-level learning in CNNs. In: Shen, D., et al. (eds.) MICCAI 2019. LNCS, vol. 11767, pp. 298–306. Springer, Cham (2019). https://doi.org/10.1007/978-3-030-32251-9_33
21. Jahanifar, M., et al.: Segmentation of lesions in dermoscopy images using saliency map and contour propagation. arXiv (2017)
22. Simonyan, K., Zisserman, A.: Very deep convolutional networks for large-scale image recognition. In: ICLR (2015)
23. Selvaraju, R.R., et al.: Grad-CAM: visual explanations from deep networks via gradient-based localization. Int. J. Comput. Vis. **128**(2), 336–359 (2019)
24. Armato III, S.G., Petrick, N.A.: Computer-aided diagnosis. In: SPIE Proceedings, vol. 10134. International Society for Optics and Photonics, Bellingham (2017). 1013428
25. K. Simonyan, A. Vedaldi, and A. Zisserman. Deep inside convolutional networks: Visualising image classification models and saliency maps. CoRR, abs/1312.6034, 2013. 3
26. Lundberg, S.M., Lee, S.I.: A unified approach to interpreting model classifications. In: Proceedings of the Advances in Neural Information Processing Systems, pp. 4768–4777 (2017)
27. Ribeiro, M.T., Singh, S., Guestrin, C.: Why should i trust you?: Explaining the classifications of any classifier. In: Proceedings of the 22nd ACM SIGKDD International Conference on Knowledge Discovery and Data Mining, pp. 1135–1144. ACM (2016)
28. Litjens, G., Debats, O., Barentsz, J., Karssemeijer, N., Huisman, H.: PROSTATEx challenge data. The Cancer Imaging Archive (2017)

DMAH 2020: NLP Based Learning from Unstructured Data

Tracing State-Level Obesity Prevalence from Sentence Embeddings of Tweets: A Feasibility Study

Xiaoyi Zhang[1] , Rodoniki Athanasiadou[2] , and Narges Razavian[2(✉)]

[1] Department of Biomedical Informatics and Medical Education,
School of Medicine, University of Washington, Seattle, WA 98195, USA
xyzhang7@uw.edu
[2] Department of Population Health, School of Medicine,
New York University, New York, NY 10012, USA
{rodoniki.athanasiadou,narges.razavian}@nyulangone.org

Abstract. Twitter data has been shown broadly applicable for public health surveillance. Previous public heath studies based on Twitter data have largely relied on keyword-matching or topic models for clustering relevant tweets. However, both methods suffer from the short-length of texts and unpredictable noise that naturally occurs in user-generated contexts. In response, we introduce a deep learning approach that uses hashtags as a form of supervision and learns tweet embeddings for extracting informative textual features. In this case study, we address the specific task of estimating state-level obesity from dietary-related textual features. Our approach yields an estimation that strongly correlates the textual features to government data and outperforms the keyword-matching baseline. The results also demonstrate the potential of discovering risk factors using the textual features. This method is general-purpose and can be applied to a wide range of Twitter-based public health studies.

Keywords: Public health informatics · Deep learning · Natural language processing · Social media

1 Introduction

Twitter, or social media in general, is a vast space for users to express opinions and sentiments. The proliferation of social media networks accelerates the generation of public health data at an unprecedented rate, allowing big data computing approaches to achieve innovative and impactful research in health sciences. Since Paul and Dredze [10] proposed the use of Twitter data for public health informatics, several studies [2,8,9,13,17] discovered strong correlations

---

This work was conducted during the first author's research intern at NYU Center for Data Science.

© Springer Nature Switzerland AG 2021
V. Gadepally et al. (Eds.): Poly 2020/DMAH 2020, LNCS 12633, pp. 141–150, 2021.
https://doi.org/10.1007/978-3-030-71055-2_12

between government statistics for specific diseases and tweets on specific topics. These studies suggest that a large number of relevant tweets can provide insight into the general health of a population.

In a recent literature review, Jordan et al. [7] showed the majority of Twitter studies in the last decade on public health informatics has relied on keywords for classification or clustering. However, since tweets contain unpredictable noises such as slang, emoji, and misspellings, the tweets retrieved from keyword-matching models often exclude relevant or include irrelevant messages. These uncertain semantics of retrieved tweets lead to unreliable estimations for public health metrics [5, 7]. Meanwhile, other studies use Latent Dirichlet allocation to learn latent topics [11, 12]. These models depend on reliable word co-occurrence statistics and typically suffer from data sparsity when applied to short documents like tweets [14].

We propose using sentence embeddings by supervised deep learning methods to overcome this shortcoming. Hashtags, the user-annotated label that clusters tweets with shared topics regardless of the diverse textual patterns, provide a natural supervision for training distributed representations of tweets. In this work, we adapt *TagSpace* [15], a convolutional neural network (CNN) that learns word and tweet embeddings in the same vector space using hashtags as supervised signals. In order to demonstrate the feasibility of this method, we address the specific task of estimating state-level obesity from tweets characterizing actual dietary habits. We use both embeddings to cluster and to extract relevant textual features that correspond to population-level dietary habits from over two hundred million tweets. The regression on these textual features strongly correlates to state-level obesity prevalence surveyed by Centers for Disease Control and Prevention [1]. Since our method is not specifically tailored to obesity research, our approach is applicable to a wide range of public health studies that involve Twitter data.

2   Data Acquisition and Pre-processing

We retrieve 272.8 million tweet records posted in 2014 using the Twitter API [1], and we assign one state among the contiguous United States (48 states plus the District of Columbia) to each of the 261 million records based on user geolocation metadata. Non-English posts are removed from our dataset. We use regex to perform the following steps:

1. Convert all alphabetical characters to lower case
2. Remove all URLs, user mentions, and special characters except the hashtag symbol #
3. Remove numerical characters except those in hashtags
4. Add white space between consecutive emojis, and limit repeating mentions of words or emojis in a post

---

[1] https://developer.twitter.com/.

After initial preprocessing, we find 9.5 million unique vocabularies (including hashtags and emojis) heavily tailed at scarce mentions - 6.3 million are mentioned only once, and 8.9 million less than ten times. Similar distribution is also discovered in hashtags. Including scarcely-mentioned words not only results in a memory-demanding lookup table, but also puts our model under the risk of overfitting, as the model may memorize the rare textual patterns found only in the training set. Hence we select 500 k most mentioned words (excluding stop words) and 50k hashtags for our model, and all out-of-vocabulary words are tokenized as <UNKNOWN>. The data pre-processing pipeline can be visualized in Fig. 1.

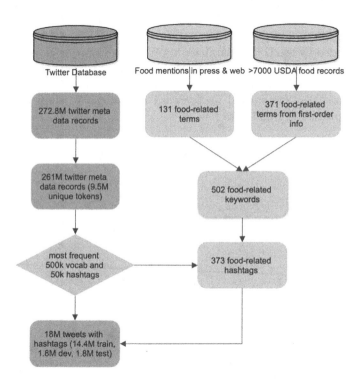

Fig. 1. Twitter data pre-processing and keyword acquisition pipeline.

3   Methods

We address the specific task of using the twitter data within a state to estimate the obesity prevalence in that state. Inspired by recent Twitter-derived public health studies [2,8,9,17], we first compile a set of keywords related to dietary habits to form a feature space in a regression scenario. We then adapt two deep learning models to retrieve food-related tweets by the scoring between embeddings of tweets and keywords. Embeddings of food-related tweets within a state will be aggregated for extracting features later used in regression.

3.1 Constructing Feature Space from Keywords

Following prior works by Nguyen et al. [9], we generate the keyword list from two sources: 1) the U.S. Department of Agriculture's National Nutrient Database [1] - from over 7000 food records found in the USDA database, we extract only the first-level information (e.g. "strawberry yogurt" and "nonfat yogurt" are both recorded as "yogurt"), which gives us 371 terms; and 2) popular food-related mentions in the press [2] - we add food (e.g. "sashimi" and "kimchi"), food-related slangs (e.g. "blt"), and chain restaurants (e.g. KFC and Starbucks) that are not included by the USDA database but frequently appear in user-generated contexts, which results in 131 additional terms. All of the 502 keywords in the list are reduced to their singular forms by NLTK lemmatizer [3], and words in the tweets are also lemmatized for keyword matching. We show the keyword acquisition process in Fig. 1.

3.2 Retrieving Relevant Tweets via Deep Learning

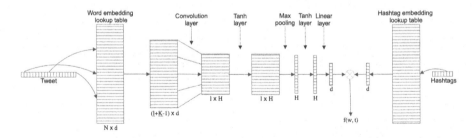

Fig. 2. The model architecture of *TagSpace* [15]. Given an input tweet w and its hashtag t, the forward pass outputs the scoring $f(w, t)$, where N denotes vocabulary size, d denotes embedding dimension, l denotes number of words (i.e. max sequence length) of input tweet, K denotes convolution window size, H denotes hidden dimension.

Keyword Matching. The simple baseline regards tweets explicitly mentioning at the one of the keywords as relevant. We find the relative frequency of food-related tweets ranges from 3.0% to 6.2% across all states with the mean value of 4.7%, which is close to statistics reported by Nguyen et al. [9] and Ghosh and Guha [6]. This implies that our pre-processing pipeline, which set a higher bar for word frequency than [9] and [6], does not drastically change the distribution of food-related tweets.

---

[2] App Spring Inc. List Challenges: Food, https://www.listchallenges.com/lists/food.

TagSpace. Simple keyword matching results in questionable semantic relevancy of retrieved tweets (e.g. "that problem is a hard *nut* to crack", "taylor swift is the *cream* of the crop"). In contrast, hashtags provide the user's labeling of a tweet's themes. We adapt *TagSpace*, a CNN model that learns the distributed representations of both words and tweets using hashtags as a supervised signal [15]. Given a tweet, the model convolves its unigrams' embeddings as input, and ranks the scoring (e.g. inner product) between the learnt embeddings of the tweet and candidate hashtags. The ranking is optimized by a pairwise hinge objective function as given by Algorithm 1, which is optimized for retrieving top-ranked hashtags according to Weston et al. [15]. We show the model architecture in Fig. 2. In our study, we consider a tweet with top ranked hashtags containing the keywords as food-related. Among the 50k hashtag candidates, 373 are included in our food keyword list. For this training task, we require the input tweets to mention hashtags in the candidate pool for prediction, and this gives us 14.4 million tweets for training, 1.8 million held out for hyperparameter tuning and 1.8 million for testing. To prevent data leakage, hashtags in the input texts are substituted with the corresponding plain words.

Algorithm 1: WARP ranking loss of *TagSpace*

Data: $e_{conv}(w) \in \mathbb{R}^{N \times d}$: sentence embedding of tweets w; $e(t) \in \mathbb{R}^d$: word embedding of hashtag t; T: set of all hashtags in corpus; T^+: set of hashtags found in w; $m \in \mathbb{R}$: margin; M: max sample iterations

Result: Optimized CNN weights, $e_{conv}(w)$, and embedding of words in w

Sample $t^+ \in \mathbb{R}^n$ from T^+

Compute $f(w, t^+) = e_{conv}(w) \cdot e(t^+)$

Sample t^- from $T \backslash T^+$

Compute $f(w, t^-) = e_{conv}(w) \cdot e(t^-)$

Initialize $i \leftarrow 0$

while $f(t^-, w) \leq m + f(t^+, w)$ *and* $i \leq M$ **do**

   | Resample t^- from $T \backslash T^+$;

   | Compute $f(w, t^-) = e_{conv}(w) \cdot e(t^-)$;

   | $i \leftarrow i + 1$

end

Compute $Loss = max(0, m - f(w, t^+) + f(w, t^-))$

Backward propagation on *Loss*

Binary TagSpace. While Weston et al. [15] optimizes the prediction of $p(hashtag \mid tweet)$ over all hashtag candidates, we are only interested in the tweets' semantic relevancy with food (i.e. $p(hashtags\ about\ food \mid tweet)$). Based on whether a tweet contains hashtags found in our food keyword list, we label all tweets with hashtags as either food-related or not. The word and tweet embeddings learnt from the CNN discussed in the previous method are optimized for a binary classification objective function instead. We use the same training and testing sets as those of the previous method.

3.3   Feature Engineering and Obesity Estimation by Elastic Net

For a given state, features are calculated from the scoring (e.g. inner product or cosine similarity) between the keyword embeddings and the average sentence-level embedding of food-related tweets within that state. The scoring function is the same for both CNN models. Both CNNs' objective functions internally train tweet embeddings in the word vector space [15,16], and hence the scoring provides information about the semantic relevance between the tweets and the keywords. By aggregating food-related tweets (i.e. tweets with sentence embeddings that have a high score with keyword vectors) within a state, we represent the dietary characteristics of that state in the word vector space. For obesity prevalence estimation, we apply the elastic net, a regression method that combines L1 and L2 regularization and has been shown to surpass ridge or lasso regressions in text regression tasks [17]. In particular, given the regression task

$$y_s = \mathbf{w}^T \mathbf{x}_s + \beta + \varepsilon$$

where $y_s \in \mathbb{R}$ denotes the obesity prevalence of a given state s, $x_s \in \mathbb{R}^{373}$ the vector of extracted textual features of state s, $\beta \in \mathbb{R}$ the intercept, and $\varepsilon \in \mathbb{R}$ an independent and zero-centered noise, the weight vector w is learnt by optimizing the objective function

$$\arg\min_{\mathbf{w},\beta} \left(\sum_{s \in S} (\mathbf{w}^T \mathbf{x}_s + \beta + \varepsilon - y_s)^2 + \lambda_1 \sum_{k=1}^{m} |w_k| + \lambda_2 \sum_{k=1}^{m} |w_k|^2 \right)$$

where λ_1 and λ_2 are the regularization coefficients, and in practice they are chosen by random search in the range $[1e{-}5, 1e2]$. We randomly hold out four states for validation and eight states for testing, and apply cross-validations for training.

4   Results and Discussion

4.1   Deep Learning Model Performance

We show the performance of two deep learning models in Table 1 based on their objective functions. Table 1a evaluates the ranking performance of our adaptation of *TagSpace*, and the result is comparable to the implementation by Weston et al. [15] on less noisy text data, which yield 37.42% P@1 and 43.01% R@10. This implies that *TagSpace* maintains its ability to predict hashtags on short and noisy documents and hence applicable to Twitter texts in general. As for the binary version of *TagSpace* shown in Table 1b, there is no prior studies for comparison. The low recall can be explained by the unbalanced labels, as in average only 9.4% of tweets in the test set contain food-related hashtags. The

Table 1. Performance of CNNs on the test set

(a) Ranking by *TagSpace*

Embedding dim	Precision@1	Recall@10
64	15.37%	40.13%
128	28.39%	43.97%
256	**32.72%**	**45.65%**

(b) Classification by *Binary TagSpace*

Embedding dim	Precision	Recall
64	73.05%	53.24%
128	84.35%	62.14%
256	**87.48%**	**66.37%**

precision of binary *TagSpace* is high, and hence we suspect if the model optimizes objective function by over-generalizing hashtag predictions (i.e. tagging tweets with only general and frequent hashtags such as #restaurant and #diet). As the model internally learns tweet embeddings, we use them to rank hashtags and find that the most frequent 100 food-related hashtags in the prediction account for 11.3% of the food-related tweets. This implies that binary *TagSpace* gives more granular information about a tweet than whether food-related or not.

4.2 Estimating Obesity Prevalence by Tweet Embeddings

Table 2. Regressions on obesity prevalence by extracting features from word and tweet embeddings

Model	dim	MAE	Pearson Corr.
Bag-of-Words	-	2.596	0.607
TagSpace	64	1.653	0.795
	128	1.571	0.813
	256	1.452	0.836
Binary TagSpace	64	1.239	0.871
	128	1.018	0.904
	256	**0.839**	**0.927**

We evaluate the regression results using mean absolute error (MAE) and Pearson correlation with government obesity data. Since no prior study has used our dataset, we handcraft a *Bag-of-Words* baseline that uses tweets filtered by *keyword matching* method and extracts features by frequencies of keywords mentioned within a state. The BOW approach is used in previous Twitter-derived obesity research [8,9]. The regression results by our baseline moderately correlates to government data, which agrees with prior works that that dietary characteristics mined from Twitter data is informative in actual obesity estimation [2,8,9]. Both CNNs generate features resulting in more accurate estimation of state-level obesity compared to our baseline, and binary *TagSpace* outperforms all other methods. Hence we are optimistic that word and tweet embeddings trained from *TagSpace* models optimizing for selective topics results in better indicators of specific diseases (Table 2).

4.3   Discovering Dietary Risk Factors with Obesity

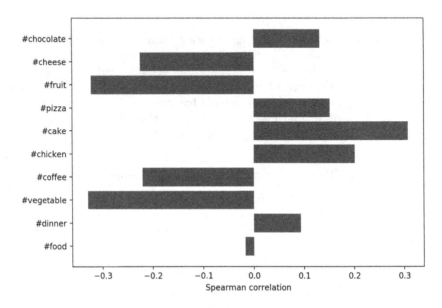

Fig. 3. Spearman correlations between selected features and obesity prevalence

We are interested in features correlating to higher obesity prevalence, and we obtain such features using Spearman correlation, which quantifies monotonic relationship between two variables. The highest positive correlation to obesity prevalence is given by ``#macncheese'' ($corr = 0.4910$), ``#wendys'' (0.4853), ``#doughnut'' (0.4796), ``#blt'' (0.4359), and ``#dominospizza'' (0.4307). We also observe that more general and frequently-mentioned features (such as #dinner, #food) usually have weaker monotonic relationship with our target variable as shown in Fig. 3. While the correlation values are modest, it points to a possibility to learn behavior risk factors from Twitter data using sentence-level embeddings of tweets.

5   Conclusion

In conclusion, we propose a deep learning approach to extract textual features using sentence-level embeddings of tweets for public health monitoring. In the case study, our adaptation of the two CNNs performs reliably on Twitter data and provides informative textual features for obesity prevalence estimation. We have also shown that features constructed via word and tweet embeddings can potentially learn risk factors for specific diseases, which is useful for monitoring acute public health incidents such as influenza tracking, allergy ailments, and infectious diseases [10,11,17]. Our data acquisition and deep learning methods do not include any obesity-related settings, which implies that our approach can be applied to a wide range of Twitter-based public health studies and for various purposes. One limitation of our study is that the demographics of Twitter users over-represent younger age-groups, and one remedy is to standardize tweets based on user ages inferred from probabilistic models [4] for future study. This work serves to inspire future studies to explore the potential of using sentence-level embeddings of social media texts for a wide scope of public health surveillance.

References

1. United states department of agriculture. national nutrient database (2014). http://ndb.nal.usda.gov/ndb/search/list?format=&count=&max=25&sort=& fg=&man=&lfacet=&qlookup=&offset=50
2. Abbar, S., Mejova, Y., Weber, I.: You tweet what you eat: studying food consumption through twitter. In: Proceedings of the 33rd Annual ACM Conference on Human Factors in Computing Systems, pp. 3197–3206. ACM (2015)
3. Bird, S., Klein, E., Loper, E.: Natural language processing with python, July 2009
4. Chamberlain, B.P., Humby, C., Deisenroth, M.P.: Probabilistic inference of Twitter users' age based on what they follow. In: Altun, Y., et al. (eds.) ECML PKDD 2017, Part III. LNCS (LNAI), vol. 10536, pp. 191–203. Springer, Cham (2017). https://doi.org/10.1007/978-3-319-71273-4_16
5. Culotta, A.: Towards detecting influenza epidemics by analyzing Twitter messages. In: Proceedings of the First Workshop on Social Media Analytics, pp. 115–122. ACM (2010)
6. Ghosh, D., Guha, R.: What are we 'tweeting'about obesity? Mapping tweets with topic modeling and geographic information system. Cartogr. Geogr. Inf. Sci. **40**(2), 90–102 (2013)
7. Jordan, S., Hovet, S., Fung, I., Liang, H., King-Wa, F., Tse, Z.: Using Twitter for public health surveillance from monitoring and prediction to public response. Data **4**(1), 6 (2019)
8. Nguyen, Q.C., et al.: Building a national neighborhood dataset from geotagged twitter data for indicators of happiness, diet, and physical activity. JMIR Public Health Surveill. **2**(2), e158 (2016)
9. Nguyen, Q.C., et al.: Twitter-derived neighborhood characteristics associated with obesity and diabetes. Sci. Rep. **7**(1), 16425 (2017)
10. Paul, M.J., Dredze, M.: You are what you tweet: analyzing twitter for public health. In: Fifth International AAAI Conference on Weblogs and Social Media (2011)
11. Paul, M.J., Dredze, M.: Discovering health topics in social media using topic models. PLoS ONE **9**(8), e103408 (2014)
12. Prier, K.W., Smith, M.S., Giraud-Carrier, C., Hanson, C.L.: Identifying health-related topics on Twitter. In: Salerno, J., Yang, S.J., Nau, D., Chai, S.-K. (eds.) SBP 2011. LNCS, vol. 6589, pp. 18–25. Springer, Heidelberg (2011). https://doi. org/10.1007/978-3-642-19656-0_4

13. Sarma, K.V., Spiegel, B.M.R., Reid, M.W., Chen, S., Merchant, R.M., Seltzer, E., Arnold, C.W.: Estimating the health-related quality of life of twitter users using semantic processing. Stud. Health Technol. Inf. **264**, 1065–1069 (2019)

14. Sridhar, V.K.R.: Unsupervised topic modeling for short texts using distributed representations of words. In: Proceedings of the 1st workshop on vector space modeling for natural language processing, pp. 192–200 (2015)

15. Weston, J., Chopra, S., Adams, K.: # tagspace: semantic embeddings from hashtags. In Proceedings of the 2014 Conference on Empirical Methods in Natural Language Processing (EMNLP), pp. 1822–1827 (2014)

16. Wu, L.Y., Fisch, A., Chopra, S., Adams, K., Bordes, A., Weston, J.: Starspace: embed all the things! In: Thirty-Second AAAI Conference on Artificial Intelligence (2018)

17. Zou, B., Lampos, V., Gorton, R., Cox, I.J.: On infectious intestinal disease surveillance using social media content. In: Proceedings of the 6th International Conference on Digital Health Conference, pp. 157–161. ACM (2016)

Enhancing Medical Word Sense Inventories Using Word Sense Induction: A Preliminary Study

Qifei Dong[1] and Yue Wang[2]([⊠])

[1] Department of Biomedical Informatics and Medical Education,
University of Washington, Seattle, WA 98195, USA
qfdong@uw.edu
[2] School of Information and Library Science, University of North Carolina
at Chapel Hill, Chapel Hill, NC 27599, USA
wangyue@email.unc.edu

Abstract. Correctly interpreting an ambiguous word in a given context is a critical step for medical natural language processing tasks. Medical word sense disambiguation assumes that all meanings (senses) of an ambiguous word are predetermined in a sense inventory. However, the sense inventory sometimes does not cover all senses or is outdated as new concepts arise in the practice of medicine. Obtaining all word senses is therefore the prerequisite work for word sense disambiguation. A classical method for word sense induction is string expansion, a rule-based method that searches the corpus for full forms of an abbreviation or acronym. Yet, it cannot be applied to ambiguous words that are not abbreviations. In this paper, we study methods that can semi-automatically discover word senses from a large-scale medical corpus, regardless of whether the word is an abbreviation. We conducted a comparative evaluation of four unsupervised data-driven methods, including context clustering, two types of word clustering, and sparse coding in word vector space. Overall, sparse coding outperforms the other methods. This demonstrates the feasibility of using sparse coding to discover more complete word senses. By comparing the senses discovered by sparse coding with those in senses inventory, we observed new word senses. For more than half of the ambiguous words in the MSH WSD data set (sense inventory maintained by National Library of Medicine), sparse coding detected more than one new word sense. This result shows an opportunity in enhancing medical word sense inventories with unsupervised data-driven methods.

Keywords: Medical word sense induction · Context clustering · Word clustering · Sparse coding

1 Introduction

Biomedical literature and clinical documents contain many ambiguous terms. For example, the word "mole" can represent a unit of amount of substance, a

© Springer Nature Switzerland AG 2021
V. Gadepally et al. (Eds.): Poly 2020/DMAH 2020, LNCS 12633, pp. 151–167, 2021.
https://doi.org/10.1007/978-3-030-71055-2_13

skin condition (nevus), and a type of mammal. The abbreviation "PCA" can mean principal component analysis, patient-controlled analgesia, and prostate cancer, among many other meanings. Automatically assigning the correct meaning (a.k.a. sense) to an ambiguous word in a context is referred to as word sense disambiguation (WSD). WSD has received extensive research in the medical domain [5,10,18,21], as it is an important step towards high-quality analysis of massive biomedical literature and clinical notes.

The first and foremost question for medical WSD is how to get all possible senses of an ambiguous word. Previous work either assumed that an existing knowledge base could provide these senses [9], or relied on human experts annotating many instances to obtain all possible senses [13]. However, the sense inventories generated from existing knowledge base sometimes do not cover all senses used in practice [9], and manual sense annotation requires specialized expertise and is time-consuming [13]. Limited amount of research also explored semi-automated approaches to discovering word senses from text corpus, i.e., word sense induction (WSI), which can discover diverse word senses with low annotation cost [22].

Two families of methods are usually used for WSI: data-driven and rule-based.

Three types of data-driven methods are commonly used. The first uses word contexts to obtain word senses. One way of using word contexts is context clustering. It starts by generating context vectors, each representing one instance of the target word's surrounding words. Then the context vectors are clustered into multiple groups, each representing a word sense. Such idea was first proposed by Schütze [19], who constructed the context vectors from second-order co-occurrence information. Researchers later employed the same idea with different context vector construction and clustering techniques [15,17,22,23]. Purandare and Pedersen [15] proposed six context clustering systems, each using a different way of constructing and clustering context vectors. In the medical domain, Xu et al. [22,23] used Expectation Maximization and Farthest First algorithms to cluster contexts. Savova et al. [17] applied the six context clustering systems proposed by Purandare and Pedersen [15] to the medical domain. Besides context clustering, some researchers created probabilistic models of the target word and its contexts. Brody and Lapata [4] proposed a Bayesian approach related to Latent Dirichlet Allocation and a generative model to find word senses. The third way of using contexts assumes the context's syntagmatic patterns are associated with the word senses. Then the main job is to find the context's syntagmatic patterns. Pustejovsky et al. [16] used Corpus Pattern Analysis to acquire the context patterns.

The second type of the data-driven methods is implemented on the word vector space. One common algorithm is to cluster the semantically similar words in the word vector space into the same group. To calculate semantic similarities, Pantel and Lin [14] employed pointwise mutual information. More recently, Arora et al. [2] studied the linear algebraic patterns in word embeddings and used sparse coding to obtain word senses.

Third, data-driven WSI can be conducted by graph-based methods. There, a graph of words is constructed and graph clustering algorithms are used to find word senses [1,7].

One common algorithm for rule-based WSI is string expansion [20], which is mainly employed in abbreviation and acronym sense induction. Each of the abbreviation's senses can be represented by one of the abbreviation's full forms. To find the full forms, some rules (regular expressions) are set to expand the abbreviation. Then these regular expressions are used to match the abbreviation's full forms in corpora. Although the rule-based method has good precision and recall [20], this method is limited to finding word senses of abbreviations and acronyms.

In this study, we aimed to study WSI regardless of whether the word is an abbreviation/acronym. We therefore focused on the unsupervised data-driven methods for WSI. We conducted preliminary study on four unsupervised data-driven methods, including context clustering, two types of word clustering, and sparse coding. We adopted a novel evaluation method that measures the overlap of sense groups without human annotation. This allowed us to efficiently compare a wide range of the unsupervised data-driven methods. The result shows sparse coding outperforms the other methods, demonstrating the feasibility of using sparse coding to discover more complete word senses.

To evaluate the potential of the unsupervised data-driven methods in enhancing existing medical word sense inventories, we manually annotated the discovered senses. We took two well-established sense inventories: 1) the test collection, MSH WSD data set [9], derived from the Unified Medical Language System, Medical Subject Headings (MeSH), and MEDLINE abstracts; and 2) the clinical abbreviation inventory from the University of Minnesota [13]. The senses of each ambiguous word were discovered by running one of the unsupervised data-driven methods on a large-scale raw text corpus. We then compared the WSI-discovered senses against existing senses in the sense inventory. A systematic analysis on the WSI-discovered senses shows that in the MSH WSD data set, half of the ambiguous words are missing more than one major senses. This analysis demonstrates the unsupervised data-driven methods have great potential in enhancing existing sense inventories by finding new senses and associated contexts.

2   Method

In this section, we discuss the data sets, the unsupervised data-driven methods for WSI, the evaluation method, and how to interpret WSI-discovered senses. The overall workflow of our study is shown in Fig. 1. Given an ambiguous word in a sense inventory and a large corpus, an unsupervised data-driven method for WSI discovered a set of senses. Then we compared the WSI-discovered word senses with the existing senses in the inventory, evaluated their overlap, and examined the new senses found by the unsupervised data-driven methods.

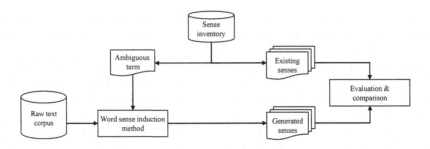

Fig. 1. Overall workflow.

To facilitate the evaluation and comparison, we represented a word sense as *a small set of semantically related words* and the set is called a *sense profile*. For example, one sense of the acronym "AB" is "AB influenza type," and its sense profile consists of a list of related words like "flu", "influenza", and "seasonal." This representation has several advantages over a natural language definition or an entry in a knowledge base. First, it is a flexible and faithful way of representing a sense, as a word shall be known "by the company it keeps [8]." Second, it does not assume the existence of a standardized superset of senses, which is why we set out finding new senses in the first place. Third, words in the sense profile can appear across corpora, making it possible to automatically and approximately estimate the overlap between senses learned from different corpora without resorting to expensive manual annotation. In our research, we set the number of words in each sense profile to 20.

2.1 Data Sets

We conducted our study on two genres of medical text: biomedical literature and clinical notes. Table 1 summarizes the basic information of the sense inventory and raw text corpus based on each of the genres. In this table, a context instance is defined by 20 words surrounding the ambiguous term (ten words each side).

Table 1. Statistics of the data sets.

		Biomedical literature	Clinical notes
Sense inventory	Name	MSH WSD data set	UMN clinical abbreviations
	# Ambiguous terms	184	75
	Avg. # context instances per ambiguous term	190	500
Raw text corpus	Name	MEDLINE abstracts	MIMIC-III clinical notes
	File size (Gigabytes)	13	4.6
	Avg. # context instances per ambiguous term	31,148	10,898

The sense inventories provide ambiguous terms, as well as their word senses and context instances. In each of these context instances, the corresponding term's word sense is known beforehand. Then for each word sense stored in the sense inventories, we could gather the context instances corresponding to this word sense. Using these gathered context instances, the sense profile of the word sense could be obtained. First, we computed the mutual information $I(s; w)$ between the sense s and any word w that appears in the gathered context instances. Then the 20 words with the highest $I(s; w)$ were selected to form the sense profile.

The sense inventories in the biomedical literature and the clinical note settings are the MSH WSD data set [9] and the University of Minnesota clinical abbreviation and acronym sense inventory (UMN) [13], respectively.

MSH WSD data set contains 203 ambiguous terms. In our experiments, we used 184 ambiguous terms. First, as our focus was single-word terms in this study, we excluded the multi-word terms. Including multi-word terms is our future direction. Second, the remaining terms with less than 100 context instances were excluded. The reason is to ensure robust estimation of mutual information $I(s; w)$ between a sense and a word. The average number of the context instances per ambiguous term is 190.

UMN contains 440 ambiguous terms. Each term is a single word. We excluded the terms with less than 100 context instances. 75 terms remained. The average number of the context instances per term is 500.

The raw text corpora are the data sets from which the unsupervised data-driven methods learn word senses of ambiguous terms. The raw text corpora in the biomedical literature and the clinical note settings are the MEDLINE abstracts and the admission notes in MIMIC-III, respectively. The average numbers of the context instances per term in the MEDLINE abstracts and the MIMIC-III clinical notes are 31,148 and 10,898, respectively.

2.2   Unsupervised Data-Driven Methods for WSI

This section describes the four unsupervised data-driven methods for WSI. Given an ambiguous word and a large corpus, the goal of the methods is to discover a set of senses in this corpus, with each sense represented by a sense profile. Among the four methods, sparse coding and the two word clustering methods require dense word vectors. We trained 100-dimensional dense word vectors using the skip-gram algorithm [11] in Google's word2vec package for single words in a case-insensitive manner. In the biomedical literature setting, we trained word vectors using MEDLINE abstracts. In the clinical note setting, we trained word vectors using MIMIC-III clinical notes.

Each of the unsupervised data-driven methods contains hyper-parameters. As a preliminary study, we intuitively explored how to select the hyper-parameters' values. A comprehensive sensitivity analysis is our future direction.

Context Clustering. The intuition of context clustering is that words with similar contexts are semantically similar to each other [12]. First, we extracted context windows with ten words on both sides of an ambiguous word in the corpus. The last row of Table 1 shows that such context windows are abundant. After finding all context windows of the word, we performed *tf-idf* weighting to obtain sparse context vectors. Then we ran k-means to cluster these sparse vectors. Each resulting cluster formed one word sense. k is a hyper-parameter, determining the number of senses we expected the method to find for each ambiguous word. In this and the following methods that use k-means, k has the same meaning. We evaluated different settings of k in the experiments (Sect. 3). To get the sense profile of each sense, we calculated the centroid of the corresponding cluster and selected the 20 words with the highest weights in the centroid vector.

Word Clustering I (Nearest Words in Context Windows). According to the distributional hypothesis [8], words tend to have related senses if they occur in similar context. First, we extracted context windows in the same way as the previous method. Then we took out all words appearing in these context windows, except the target word itself and the stopwords like "the" and "of."[1] Each extracted word could be represented by a dense word vector. We ran k-means on these dense word vectors and obtained k centroids. The 20 words whose dense word vectors were closest to the centroid were used as the words in the sense profile. The distance between two vectors in the word vector space was measured by cosine distance.

Word Clustering II (Nearest Words in Vector Space). This algorithm obtains word senses directly in word vector space. As nearby words in word vector space are semantically related to each other, the senses of a target word could be contained in its neighbors. For the same reason, these neighboring words cannot distribute evenly in space. Instead, they should cluster into groups to form senses.

To get the k senses of a target word, we ran k-means on the target word's N nearest neighbors in word vector space. Then for each sense, we selected 20 words whose dense word vectors are closest to the corresponding cluster centroid as the words in the sense profile. Note an appropriate value of N is important. If N is too small, we could miss certain senses. If N is too large, we could obtain irrelevant senses. We empirically set $N = 500$ as a reasonable size.

Sparse Coding. We adopted the sparse coding method from Arora *et al.*'s work [2]. Sparse coding works directly in word vector space, and is based on the assumption that each word is a linear combination of some word cluster centroids [2]. Each of these clusters contains several words and forms a sense. For example, the word *BAT* could mean *Chiroptera*, a kind of mammal, as well as *Brown Fat*. Let d denote the number of dimensions of the word vectors,

---

[1] The full list of stopwords is available at https://www.ranks.nl/stopwords.

$v_{BAT} \in \mathbb{R}^d$ denote the word vector of BAT, and $c_{BAT1}, c_{BAT2} \in \mathbb{R}^d$ denote the cluster centroids representing the two senses of BAT, respectively. Ideally, $v_{BAT} = r_{BAT1} c_{BAT1} + r_{BAT2} c_{BAT2}$, where r_{BAT1} and r_{BAT2} are coefficients.

Let m denote the total number of word clusters in the whole word vector space. m needs to be pre-defined before conducting sparse coding. We set $m = 2,000$ as suggested in the previous work [2]. Let $\{c_i \in \mathbb{R}^d\}_{i=1}^m$ denote the set of the word cluster centroids. Let V denote the vocabulary, i.e., the set of all words in a corpus. When representing the word $w_j \in V$, we use $r_{j,i}$ to denote the coefficients multiplying c_i. Let $\epsilon_j \in \mathbb{R}^d$ denote a noise vector. Then for any word $w_j \in V$, its word vector can be represented by $\sum_{i=1}^m r_{j,i} c_i + \epsilon_j$. As a word can only have a small number of senses, most of the coefficients $r_{j,i}$ should be zero, hence the name *sparse* coding. Each of the cluster centroids multiplied by a non-zero coefficient represents one of the target word's senses. Our goal is to 1) find all of the word clusters in the word vector space, and 2) for each target word, obtain the cluster centroids with non-zero coefficients. The number of the cluster centroids with non-zero coefficients is a pre-defined hyper-parameter and denoted by k. k determines the number of senses we expected sparse coding to find for each ambiguous word. This k intrinsically has the same meaning as k in k-means mentioned above does. When implementing the sparse coding method, we varied k for different medical text corpora. For each resulting sense, we selected 20 words whose dense word vectors are closest to the corresponding cluster centroid as the words in the sense profile.

2.3   Evaluation Method

As the senses discovered by the unsupervised data-driven methods are represented by the sense profiles rather than human readable labels, we need a method for telling whether a discovered sense matches any actual sense. We adopted the novel evaluation method called "police lineup" proposed by Arora *et al.* [2]. This evaluation method estimates the degree of overlap between two sets of senses without resorting to labeling context instances in corpora. It allows efficient evaluation of different unsupervised data-driven methods against a sense inventory. Intuitively, it tests how well the WSI-discovered senses match the actual senses and distinguish from the irrelevant ones. It is called "police lineup" because it is analogous to the investigation process where a witness has to identify the suspect from several innocent people.

We illustrate the police lineup evaluation in Fig. 2. This figure depicts a word vector space, where semantically related words are close to each other. The pentagon is the target word, associated with the vector v_t. S_1–S_6 represent senses existing in the sense inventory. S_1, S_2, and S_3 are the target word's actual senses in the sense inventory. S_4, S_5, and S_6 do not belong to the target word, and hence are termed *distracting senses*. The solid circles inside each of S_1–S_6 represent the words in the corresponding sense profile. Each of the 5-point stars c_1, c_2, and c_3 represents a WSI-discovered sense, obtained by calculating the mean of the word vectors in the corresponding sense profile. The police lineup evaluation asks each of the WSI-discovered senses to return its closest existing sense(s). In Fig. 2, c_1 picks S_1, c_2 picks S_5, and c_3 picks S_2. The picked senses are called candidate senses. Then precision and recall can be calculated. Precision is the number of the actual senses picked out divided by the number of the candidate senses. Recall is the number of the actual senses picked out divided by the total number of the actual senses of the target word.

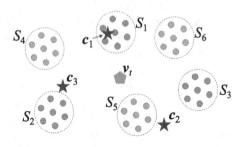

Fig. 2. The police lineup evaluation method.

In practice, we randomly selected some of the existing senses not attributed to the target word as the distracting senses. The number of the selected distracting senses should be set properly. If this number is too large, the senses discovered by either mediocre or worse WSI method could hardly pick the actual senses. As a result, we could hardly compare the performance of these methods using police lineup evaluation. If this number is too small, the senses discovered by either mediocre or better WSI method could accurately pick the actual senses. In our study, given a target word, the number of its distracting senses was set to 20 minus the number of the actual senses of the target word.

Overlap score between a WSI-discovered sense and an existing sense is used to decide which existing sense(s) the WSI-discovered sense should pick. Let C_t denote the set of WSI-discovered senses of the target word. Each element $c \in C_t$ represents a WSI-discovered sense, obtained by getting the mean of the word vectors in the corresponding sense profile. Let n denote the number of words in the sense profile. Let d denote the number of dimensions the word vectors have. Let T_t denote the set containing the actual and distracting senses. Each element $\mathbf{S} \in T_t$ is a matrix in $\mathbb{R}^{n \times d}$ representing one sense in T_t. Each row in $\mathbf{S}$ is a vector representing a word in the sense profile of this sense. The row order does not matter. Let $\|\cdot\|_2$ denote the L_2 norm of a vector. Let $|V|$ denote the number of words in the vocabulary V. We calculate the overlap score between a WSI-discovered sense c and a sense $\mathbf{S}$ for the target word using the following formula:

$$score_t(c, \mathbf{S}) = \left(\|\mathbf{S}c\|_2 - \frac{1}{|C_t|} \sum_{c' \in C_t} \|\mathbf{S}c'\|_2 \right) + \left(\|\mathbf{S}v_t\|_2 - \frac{1}{|V|} \sum_{j=1}^{|V|} \|\mathbf{S}v_j\|_2 \right).$$

$$(1)$$

The two parts measure how close c is to S and how close S is to v_t, respectively. The second part is to prevent the cases that both the WSI-discovered sense and the sense in T_t are too far away from the target word, which can happen if the sense in T_t is a distracting one. The two subtracted terms are average similarities. Algorithm 1 describes the process to pick candidate senses from the actual and distracting senses.

Algorithm 1. Police lineup evaluation

1: Initialize: an empty set used to contain candidate senses, *candidates*
2: **for** each cluster centroid $c \in C_t$ **do**
3:     **for** each sense S in the set T_t **do**
4:         Calculate the overlap score between c and S using Equation (1)
5:     Let $U := \{$top 2 highest-scoring $S\}$
6:     *candidates* $\leftarrow$ *candidates* $\cup U$
7: **return** Top p senses with the highest scores in *candidates*

By varying p, we got different precision and recall for detecting the actual senses out of the candidate senses. We varied p from 1 to the maximum number of elements that the set *candidates* could have, $2 \times |C_t|$. Then we drew the precision-recall curve. A larger area under the curve means the WSI-discovered senses cover more actual senses. As these actual senses are typical senses stored in sense inventories, finding more actual senses indicates the unsupervised data-driven method is more reliable.

2.4   Interpreting WSI-Discovered Senses

Ideally, we could interpret a WSI-discovered sense using the words in the corresponding sense profile. Yet, it can be a tough task as these words are professional and difficult for interpretation. We used two approaches to better interpret the WSI-discovered senses. First, given a word in the sense profile, we printed out its semantically related words that are more commonly used and more familiar to laymen. These common words were also added to the sense profile. We term the original 20 words in the sense profile *precise sense-profile words*, and the common ones *common sense-profile words*. Second, given one WSI-discovered sense of a target word, we extracted the context instances where the target word bears this sense from MEDLINE abstracts or MIMIC-III clinical notes.

Identifying Common Sense-Profile Words. In word vector space, the frequently used words close to the precise sense-profile words were found. These frequently used words are the common sense-profile words. Algorithm 2 describes the detail. The number of common sense-profile words returned is 20.

Algorithm 2. Finding common sense-profile words

1: **Input:**
   1) Precise sense-profile words of a sense. Let S denote the set containing the precise sense-profile words and $s \in S$ denote one of them. Let n denote the number of the precise sense-profile words in each sense profile
   2) A corpus
   3) Word vectors
2: Given the corpus, rank all words by their frequency from high to low. Take out the words from Rank 1 – 8000
3: Group these words by their ranking. Specifically, words from Rank 1 – 1000, 1001 – 2000, 2001 – 4000 and 4001 – 8000 are grouped to 4 sets, respectively denoted by G_1, G_2, G_3, and G_4
4: Initialization: Let Out denote the output set, which is initialized to an empty set
5: **for** $i = 1, 2, 3, 4$ **do**
6:    **for** each word $w \in G_i$ **do**
7:       $score(w, S) = \frac{1}{n} \sum_{s \in S} \cos(\boldsymbol{v}_w, \boldsymbol{v}_s)$
8:    Let $U := \{\text{top 5 highest-scoring } w\}$
9:    $Out \leftarrow Out \cup U$.
10: **return** Out

Extracting Representative Context Instances. Another way to interpret a WSI-discovered sense is to look at the context instances where the ambiguous word most likely to bear that sense. Hence, we need a method to automatically infer what the ambiguous word means in a certain context instance. To do this task, we used the precise sense-profile words of the WSI-discovered senses.

A context instance x consists of a set of words $\{w_i\}_{i=1}^{m}$ (excluding the target word itself), where m is the number of words in this set. $\boldsymbol{v}_i$ denotes the word vector of w_i in this set. Let Y be the set of the WSI-discovered senses. A sense $y \in Y$ is represented by a set of precise sense-profile words $\{w_j\}_{j=1}^{n}$, where n is the number of the words. $\boldsymbol{v}_j$ denotes the word vector of the precise sense-profile word w_j. The relatedness score between x and y is evaluated by

$$r(x, y) = \sum_{j=1}^{n} \max_{1 \leq i \leq m} \cos(\boldsymbol{v}_i, \boldsymbol{v}_j).$$

This formula means given any word representing the sense y, if the context instance x contains at least one word semantically similar to the given word, the context instance x is related to the sense y. The scores can be converted into a probability distribution of senses: $p(y|x) = \exp[r(x, y)] / \sum_{y' \in Y} \exp[r(x, y')]$. To interpret sense y, we can look at those context instances with the highest $p(y|x)$ and examine the sense of the ambiguous word in those contexts.

Note even with the above two approaches (showing the common sense-profile words and representative context instances), we could still occasionally fail to interpret WSI-discovered senses due to low-quality clustering results. We call such WSI-discovered senses "unclear senses."

3   Results

Our experiments ran on a MacBook Pro laptop with one four-core Intel Core i7-7700HQ 2.8GHz central processing unit, 16GB memory, one 256GB Macintosh HD disk, and running the macOS High Sierra 10.13 operating system. The context clustering method was written in Python 3.6. The other unsupervised data-driven methods and the evaluation method were written in MATLAB R2017b. Sparse coding was solved by a MATLAB toolbox called SMALLbox [6].

3.1   Precision-Recall Curves

We conducted each unsupervised data-driven method twice on each genre of the medical text (biomedical literature or clinical notes). Each time, we set k to a different value. Recall k is the number of word senses we expected a method to find for one ambiguous term. In the biomedical literature setting, we set k to 4 and 5. In the clinical note setting, k was set to 5 and 8. k was set larger in the clinical note setting because the average number of actual senses of a

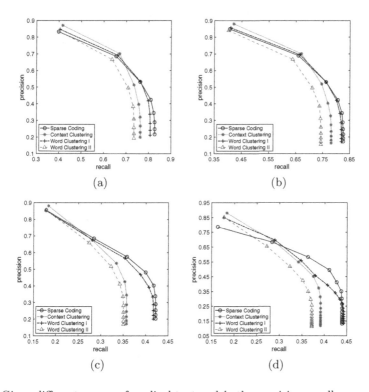

(a)   (b)

(c)   (d)

Fig. 3. Given different genres of medical text and k, the precision-recall curves for the unsupervised data-driven methods. (a) $k = 4$ in the biomedical literature setting. (b) $k = 5$ in the biomedical literature setting. (c) $k = 5$ in the clinical note setting. (d) $k = 8$ in the clinical note setting.

Table 2. Area under the precision-recall curve.

	Biomedical literature		Clinical notes	
	$k = 4$	$k = 5$	$k = 5$	$k = 8$
Sparse coding	**0.290**	**0.286**	**0.160**	**0.176**
Context clustering	0.251	0.258	0.119	0.138
Word clustering I	0.280	0.276	0.147	0.174
Word clustering II	0.231	0.234	0.113	0.122

clinical abbreviation is larger. To compare the methods, we used the police lineup evaluation to generate the precision-recall curves, shown in Fig. 3. Table 2 lists the area under each precision-recall curve.

Figure 3 shows sparse coding outperforms the other methods across all tests, especially at the high recall end. Tables 2 shows sparse coding has the largest area under the curve in each test.

3.2   Case Studies of the Word Senses Discovered by Sparse Coding

In this section, we show case studies of the word senses discovered by sparse coding on the biomedical literature setting. We provide the sense profiles and context instances of the word *EPI*. We also list some WSI-discovered senses not stored in the existing inventories.

Sense Profile. Table 3 shows a WSI-discovered sense of the word *EPI* with the sense profile. By looking at both precise and common sense-profile words, we can infer that this sense is related to hormone. By further consulting representative context instances, we know the sense is *Epinephrine*, a kind of hormone.

Table 3. A WSI-discovered sense of the word *EPI* and the sense profile (only the ten most representative precise and common sense-profile words are shown).

Precise sense-profile words	Common sense-profile words	Annotated sense
crh trh adrenocorticotropin corticotropin beta-end thyrotrophin-releasing crf-41 beta-endorphin acth corticoliberin	hormone insulin secretion gh pituitary dopamine acth prl prolactin trh	Epinephrine

Representative Context Instances. Table 4 shows context instances for another sense of *EPI*, with estimated $p(y|x)$ and PubMed Identifiers (PMID). It is clear that *EPI* means *Echo-Planar Imaging* here. This sense is not included in the MSH WSD data set.

Table 4. Example context instances associated with the sense *Echo-Planar Imaging*.

PMID	$p(y|x)$	Context instance
12417991	1.000	*... within the brain, causing geometric distortions in echo planar imaging (EPI). Even if subtle, change in shim can lead to artifactual ...*
9332249	0.984	*... each of which consists of a number of gradient echoes (EPI factor, EF). The aim of our studywas to evaluate ...*
15670684	0.960	*... brain activation can be monitored during the ongoing scan. However, EPI suffers from geometric distortions due to inhomogeneities of the magnetic ...*

We further estimated the frequency of the newly discovered sense to see whether it is a major sense. Take *EPI = echo planar imaging* as an example, there were 614 context instances in the MEDLINE abstracts with $p(y|x) \geq 0.960$. We randomly selected 50 out of the 614 context instances and found that in 48 (96%) of them, *EPI* indeed means *Echo-Planar Imaging*. Therefore, this is a major sense of *EPI* used in hundreds of the MEDLINE abstracts. We did the same trial on the newly discovered senses of other words. Most of the newly discovered senses are major senses.

Newly Discovered Senses. Sparse coding identified a large number of newly discovered senses. Table 5 shows some examples. In this table, "existing senses" are provided by the MSH WSD data set. "WSI-discovered senses" can be further divided into two groups: 1) senses included in the existing senses ("overlapping existing senses"), and 2) senses not included in the existing senses ("newly discovered"). We also documented the number of unclear senses in this table. Across all 184 ambiguous terms in the MSH WSD data set, the average sense overlap is 61.4% per word. Sparse coding found the new senses for 100 ambiguous words, and a total number of 162 new senses were found. This means more than half of the words in the MSH WSD data set miss at least one major sense (1.62 senses to be exact). Our analysis results, including the sense profiles and the senses inferred using the context instances, are available at http://bit.ly/2Id6837.

Table 5. Comparison between the senses stored in the MSH WSD data set and the WSI-discovered senses.

Ambiguous term	Existing senses	WSI-discovered senses		Number of unclear senses
		Overlapping existing senses	Newly discovered	
Epi	Epirubicin; Epinephrine	Epirubicin; Epinephrine	Echo-planar imaging; Extended program of immunization	1
Moles	Talpidae; Nevus	Talpidae; Nevus	Mole, unit of measurement; Hydatidiform mole	0

4 Discussion

Overall, sparse coding outperforms the other three methods, especially at the high recall end. This means in most cases, using sparse coding can discover more complete word senses from large-scare medical text. Sparse coding computes a global set of senses, while the word clustering methods compute senses for each word locally. Sparse coding outperforming word clustering indicates medical words may have very diverse and uncorrelated senses that do not belong to a local region of the word semantic space. Both sparse coding and word clustering I perform better than context clustering. This indicates clusters in dense word vector space could better represent a sense than clusters consisting of sparse context vectors.

Our analysis shows the MSH WSD data set can be enhanced using the unsupervised data-driven methods. The MSH WSD data set missing major senses could result from some problems occurred in the construction steps of this data set [9]. The MSH WSD data set was constructed by three steps. First, Unified Medical Language System (UMLS) [3] was screened to get ambiguous terms. Each term was linked to some MeSH terms, which served as senses. Second, each ambiguous term and its related MeSH terms were used to extract MEDLINE citations, which served as context instances. Third, the researchers eliminated trivial and repeated senses using three filters. The first filter removed the trivial senses whose corresponding context instances were very few. By conducting Support Vector Machine on the extracted MEDLINE citations of each ambiguous term, the second filter checked whether some of the term's senses were semantically similar and removed the repeated ones, if any. The third filter removed single-letter terms. In our research, the unsupervised data-driven methods discovered two types of senses not in the MSH WSD data set. One type, like *Extended Program of Immunization*, is included in neither MeSH vocabulary nor UMLS. These senses could not be obtained in the first step of constructing the MSH WSD data set. The other type is included in MeSH vocabulary or UMLS, which means such senses were removed by the filters. Yet, some newly discovered senses like *Echo-Planar Imaging* and *Mole, Unit of Measurement* are frequently used and distinct from other senses, and hence should not have been removed. Two reasons could explain why such senses were filtered out: 1) the senses were not commonly used when the MSH WSD data set was constructed, and hence removed by the first filter; and 2) in the second filter, Support Vector Machine did not give sufficiently accurate classification results.

There are several directions for future work. First, we only systematically annotated the WSI-discovered senses of the ambiguous terms in the MSH WSD data set, because it is relatively easy to understand their contexts – biomedical literature. In the future, we aim to collaborate with domain experts to annotate the WSI-discovered senses for clinical abbreviations and compare the senses discovered by WSI to those in the clinical sense inventory. Second, unsupervised learning algorithms, especially flat clustering algorithms like k-means, could sometimes generate low-quality clusters. To improve the clusters' quality, a comprehensive sensitivity analysis should be conducted to obtain proper ranges of the hyper-parameters in these algorithms. Also, we will try more powerful clustering algorithms like Tight Clustering for Rare Senses [23]. Third, the current work did not include multi-word ambiguous terms. Exploring these terms is another future direction.

5   Conclusion

In this paper, we did a preliminary study on the four unsupervised data-driven methods for WSI, including context clustering, two types of word clustering, and sparse coding. We applied these unsupervised data-driven methods on two genres of medical text, biomedical literature and clinical notes. Among the four methods, sparse coding outperforms the other three methods, showing the feasibility of using sparse coding to discover more complete word senses from large-scare medical text. We analyzed the senses discovered by the unsupervised data-driven methods against those in the existing sense inventories. Our analysis showed that the sparse coding method detected more than one major sense for more than half of the ambiguous words in the MSH WSD data set. This result demonstrates that it is very promising to employ the unsupervised data-driven methods to improve sense coverage in the existing sense inventories.

References

1. Agirre, E., Martínez, D., de Lacalle, O.L., Soroa, A.: Two graph-based algorithms for state-of-the-art WSD. In: Proceedings of the 2006 Conference on Empirical Methods in Natural Language Processing, Sydney, Australia, pp. 585–593. Association for Computational Linguistics (2006)
2. Arora, S., Li, Y., Liang, Y., Ma, T., Risteski, A.: Linear algebraic structure of word senses, with applications to polysemy. Trans. Assoc. Comput. Linguist. **6**, 483–495 (2018)
3. Bodenreider, O.: The unified medical language system (UMls): integrating biomedical terminology. Nucleic Acids Res. **32**(Suppl. 1), D267–D270 (2004)

4. Brody, S., Lapata, M.: Bayesian word sense induction. In: Proceedings of the 12th Conference of the European Chapter of the Association for Computational Linguistics, Athens, Greece, pp. 103–111. Association for Computational Linguistics (2009)
5. Chen, Y., Cao, H., Mei, Q., Zheng, K., Xu, H.: Applying active learning to supervised word sense disambiguation in MEDLINE. J. Am. Med. Inform. Assoc. **20**(5), 1001–1006 (2013)
6. Damnjanovic, I., Davies, M.E.P., Plumbley, M.D.: SMALLbox - an evaluation framework for sparse representations and dictionary learning algorithms. In: Vigneron, V., Zarzoso, V., Moreau, E., Gribonval, R., Vincent, E. (eds.) LVA/ICA 2010. LNCS, vol. 6365, pp. 418–425. Springer, Heidelberg (2010). https://doi.org/10.1007/978-3-642-15995-4_52
7. Di Marco, A., Navigli, R.: Clustering and diversifying web search results with graph-based word sense induction. Comput. Linguist. **39**(3), 709–754 (2013)
8. Firth, J.R.: A Synopsis of Linguistic Theory, 1930–1955. Studies in Linguistic Analysis (1957)
9. Jimeno-Yepes, A.J., McInnes, B.T., Aronson, A.R.: Exploiting MeSH indexing in MEDLINE to generate a data set for word sense disambiguation. BMC Bioinform. **12**(1), 223 (2011)
10. Liu, H., Teller, V., Friedman, C.: A multi-aspect comparison study of supervised word sense disambiguation. J. Am. Med. Inform. Assoc. **11**(4), 320–331 (2004)
11. Mikolov, T., Sutskever, I., Chen, K., Corrado, G.S., Dean, J.: Distributed representations of words and phrases and their compositionality. In: Proceedings of the 26th International Conference on Neural Information Processing Systems, Lake Tahoe, Nevada, USA, pp. 3111–3119. Curran Associates Inc. (2013)
12. Miller, G.A., Charles, W.G.: Contextual correlates of semantic similarity. Lang. Cogn. Process. **6**(1), 1–28 (1991)
13. Moon, S., Pakhomov, S., Liu, N., Ryan, J.O., Melton, G.B.: A sense inventory for clinical abbreviations and acronyms created using clinical notes and medical dictionary resources. J. Am. Med. Inform. Assoc. **21**(2), 299–307 (2013)
14. Pantel, P., Lin, D.: Discovering word senses from text. In: Proceedings of the 8th ACM SIGKDD International Conference on Knowledge Discovery and Data Mining, Edmonton, Canada, pp. 613–619. Association for Computing Machinery (2002)
15. Purandare, A., Pedersen, T.: Word sense discrimination by clustering contexts in vector and similarity spaces. In: Proceedings of the 8th Conference on Computational Natural Language Learning (CoNLL 2004) at HLT-NAACL 2004, Boston, MA, USA. Association for Computational Linguistics (2004)
16. Pustejovsky, J., Hanks, P., Rumshisky, A.: Automated induction of sense in context. In: Proceedings of the 20th International Conference on Computational Linguistics, Geneva, Switzerland, pp. 924–930. COLING (2004)
17. Savova, G., Pedersen, T., Purandare, A., Kulkarni, A.: Resolving ambiguities in biomedical text with unsupervised clustering approaches. University of Minnesota Supercomputing Institute Research Report (2005)
18. Schuemie, M.J., Kors, J.A., Mons, B.: Word sense disambiguation in the biomedical domain: an overview. J. Comput. Biol. **12**(5), 554–565 (2005)
19. Schütze, H.: Automatic word sense discrimination. Comput. Linguist. **24**(1), 97–123 (1998)
20. Siklósi, B., Novák, A., Prószéky, G.: Resolving abbreviations in clinical texts without pre-existing structured resources. In: 4th Workshop on Building and Evaluating Resources for Health and Biomedical Text Processing (2014)

21. Xu, H., Markatou, M., Dimova, R., Liu, H., Friedman, C.: Machine learning and word sense disambiguation in the biomedical domain: design and evaluation issues. BMC Bioinform. **7**(1), 334 (2006)
22. Xu, H., Stetson, P.D., Friedman, C.: Methods for building sense inventories of abbreviations in clinical notes. J. Am. Med. Inform. Assoc. **16**(1), 103–108 (2009)
23. Xu, H., Wu, Y., Elhadad, N., Stetson, P.D., Friedman, C.: A new clustering method for detecting rare senses of abbreviations in clinical notes. J. Biomed. Inform. **45**(6), 1075–1083 (2012)

DMAH 2020: Biomedical Data Modelling and Prediction

Teaching Analytics Medical-Data Common Sense

Tomer Sagi[1,2](✉) [iD], Nitzan Shmueli[3] [iD], Bruce Friedman[3], and Ruth Bergman[3]

[1] Department of Information Systems, University of Haifa, Haifa, Israel
tsagi@is.haifa.ac.il
[2] Department of Computer Sciences, Aalborg University, Aalborg, Denmark
[3] Acute Care, GE Healthcare, Haifa, Israel

Abstract. The availability of Electronic Medical Records (EMR) has spawned the development of analytics designed to assist caregivers in monitoring, diagnosis, and treatment of patients. The long-term adoption of these tools hinges upon caregivers' confidence in them, and subsequently, their robustness to data anomalies. Unfortunately, both complex machine-learning-based tools, which require copious amounts of data to train, and a simple trend graph presented in a patient-centered dashboard, may be sensitive to noisy data. While a caregiver would dismiss a heart rate of 2000, a medical analytic relying on it may fail or mislead its users. Developers should endow their systems with medical-data common sense to shield them from improbable values. To effectively do so, they require the ability to identify them. We motivate the need to teach analytics common sense by evaluating how anomalies impact visual-analytics, score-based sepsis-analytics SOFA and qSOFA, and a machine-learning-based sepsis predictor. We then describe the anomalous patterns designers should look for in medical data using a popular public medical research database - MIMIC-III. For each data type, we highlight methods to find these patterns. For numerical data, statistical methods are limited to high-throughput scenarios and large aggregations. Since deployed analytics monitor a single patient and must rely on a limited amount of data, rule-based methods are needed. In light of the dearth of medical guidelines to support such systems, we outline the dimensions upon which they should be defined upon.

Keywords: Data quality · Sepsis · ICU · EMR

1 Introduction

Medical research has enjoyed increased availability of large-scale medical datasets for secondary use [1], which have been used in multiple research papers (e.g., for prediction of readmission [2]). Concurrently, recent advancements in data analysis, especially in Machine Learning (ML) techniques, have brought about increased confidence in their ability to generate reliable and meaningful insights. The plethora of papers (see reviews [3,4]) demonstrating the ability of advanced

© Springer Nature Switzerland AG 2021
V. Gadepally et al. (Eds.): Poly 2020/DMAH 2020, LNCS 12633, pp. 171–187, 2021.
https://doi.org/10.1007/978-3-030-71055-2_14

analytics to predict deterioration, identify conditions, and recommend treatment, have prompted vendors to aggressively pursue new and exciting analytics, predictors, and decision support tools. All these analytics are trained and evaluated upon secondary-use clinical data. However, MIMIC-III data is noisy. Maslove et al. [5] analyzed five major vital signs and found that 15–38% of vital-sign-days contained at least one statistical outlier. Not all outliers are errors; some may be caused by patient variability. Statistical methods rely upon the availability of numerous samples from the specific patient. An assumption only valid for a few parameters, such as heart-rate and respiratory-rate, and even for these, only after a while. Stream-based methods (e.g., [6]), are similarly limited in their applicability. Assuming we may not be able to filter out errors from collected patient data reliably, how robust are simple and ML-based complex analytics to noise? In this work, we aim to perform a systematic exploration of error in EMR data in the single-patient scenario. We do not limit ourselves to those parameters for which a large volume of data exists for each patient, but rather assume a low-volume scenario as follows.

We wish to monitor all the patient's clinical measurements entered into the EMR in order to ensure that downstream analytics are warned of anomalous values. Warnings may serve to exclude readings or to highlight results based upon suspicious data. We can use the values recorded thus far for this patient, and any known statistics and limits regarding this parameter to evaluate the new value. Our patient may be being cared for in an ICU, in a ward, or even a general practitioner's office. Table 1 contains a series of hourly respiratory rates

Table 1. A series of respiratory rates from MIMIC-III

Time	Value	Run. Avg	Run. Std. Dev	Run. Lower Lim	Run. Upper Lim	Is Valid	Is Valid (Full MIMIC)	Is Valid (Grubbs)
14:00	22						TRUE	
15:00	29	22.0					TRUE	
16:00	22	25.5	4.9	15.6	35.4	TRUE	TRUE	
17:00	34	24.3	4.0	16.3	32.4	FALSE	FALSE	
18:00	50	26.8	5.9	15.0	38.5	FALSE	FALSE	
21:00	19	26.8	11.6	3.6	49.9	TRUE	TRUE	
22:00	23	21.0	11.5	−2.0	44.0	TRUE	TRUE	TRUE
23:00	11	21.3	10.8	−0.3	42.9	TRUE	TRUE	TRUE
0:00	18	20.0	11.7	−3.5	43.5	TRUE	TRUE	TRUE
1:00	15	19.8	11.3	−2.8	42.4	TRUE	TRUE	TRUE
2:00	8	19.3	11.2	−3.0	41.6	TRUE	FALSE	TRUE
3:00	12	18.3	11.7	−5.1	41.6	TRUE	TRUE	TRUE
4:00	19	17.8	11.6	−5.4	40.9	TRUE	TRUE	TRUE
5:00	6	17.8	11.1	−4.3	40.0	TRUE	FALSE	TRUE
6:01	4	17.0	11.5	−5.9	39.9	TRUE	FALSE	TRUE
7:50	22	16.1	11.8	−7.5	39.8	TRUE	TRUE	TRUE
8:00	22	16.5	11.5	−6.4	39.4	TRUE	TRUE	TRUE

reported in MIMIC-III. In a ward scenario, we would expect an even lower rate of measurement. Since no accepted medical guideline defines human limits on respiratory rate, we use our judgment to assume all values under 0, and over 100 are physically impossible and, therefore, invalid. The average and standard deviation of respiratory rate (item-id $= 220210$) over 2,726,962 such measurements in MIMIC-III are 20.12 and 5.84, respectively. Using the common outlier limit of two standard deviations, we can set reasonable bounds for respiratory rate in two manners. One is by calculating the running average and standard deviation of this measure for the specific patient, and the other is to use the general population statistics derived from the entire MIMIC-III database: [8.4,31.8]. Using the first method requires collecting at least two samples and can, thus, be applied only from the third measurement. In our example, this method highlights 34 and 50 as anomalous. Using MIMIC-generated limits flags 8, 6, and 4, which appear later in the series. Alternatively, we can use Grubbs's test [7], which requires at least six measurements and, therefore, does not judge the readings of 34 and 50. This test does not find any anomalous readings.

The statistical tests lack essential information. Deciding whether or not a measurement is anomalous must rely upon the specific medical condition of the patient, their age, and other factors. Any common sense required to judge a measurement's correctness before using it in analytics cannot be fully addressed by statistical methods in a low volume sample. Similarly, advanced machine learning methods (e.g., [8]) require a large number of observations that are unavailable in this scenario. The rest of this work is under the assumptions of this *single patient low-volume scenario*. We begin by describing the dataset and methods used (Sect. 2. We then motivate the need to teach analytics common sense by presenting an empirical analysis of the effects of anomalies on a variety of analytics (Sect. 3). Section 4 represents our main contribution, where we describe how designers can teach their analytics commons sense. Our contribution there is threefold, comprised of (1) a process template for designers to use when planning (2) a systematic review of anomalous patterns in medical data and how to identify them, and (3) a review of the facets by which common-sense rules for medical parameters should be constructed. We complete the paper by offering a brief review of related work (Sect. 5) and presenting our conclusions Sect. 6.

2   Dataset and Methods

2.1   A Systematic Analysis of Anomalous Patterns in EMR Data

MIMIC-III [9] is one of the largest publicly available sources of patient data. Researchers have developed numerous analytics (e.g., to predict sepsis [10] and choose a treatment [11]) by using MIMIC parameters which can be collected in most critical care settings. In a recent survey [12] of ML-models developed using EMR data, 22 of 98 papers reviewed used MIMIC-II/III. The MIMIC-III schema[1] contains 26 data tables of which 16 are transactional, containing

---

[1] https://mit-lcp.github.io/mimic-schema-spy/, Accessed Apr 19, 2019.

events related to a patient. The rest are dimensional, i.e., contain additional information required to perform analysis and to display human-readable results. For example, the d_items table contains the labels of parameters mentioned only by their item-id in the chartevents table. Our analysis focuses on transactional tables, predominantly chartevents (CE), with over 330 million events. CE's schema contains subject, admission, and icustay ids to identify the patient, specific hospital-stay, and ICU-stay respectively. A caregiver-id (cgid) identifies the person entering the data; charttime and storetime represent timestamps; itemid, value, valuenum, and valueuom record the parameter recorded its value, and the unit-of-measurement respectively. In the following section, we present the different anomaly types according to these fields, aligning ourselves with the type of analytics expected over this data. Identifier and timestamp anomalies affect the ability of process-based analytics (see review [13]) to collect related events and put them in sequence. Numerical anomalies affect both simple visual analytics such as trend graphs and complex machine-learning-based predictors. Textual errors affect text-mining techniques (see review [14]).

Identifiers and Timestamps. We redo the identifier completeness analysis done by Kurniati et al. [13] on a more up-to-date version of MIMIC-III where each transactional table is checked for the completeness of foreign-key codes. Under a baseline expectation of timestamps to reflect the actual time an event occurred, we expect them to be uniformly distributed throughout the hour. Any variations from this uniform distribution should reflect the inherent clinical rhythms of a hospital, such as shift changes and physician rounds. However, if the variations do not match these rhythms, analytics relying on them should be shielded accordingly. We evaluate actual timestamps against a uniform distribution and report on the variations found and their possible sources. Similarly, comparing event times for which there exist clinical/common-sense rules to govern their reasonable relation can be used to identify data entry errors. To exemplify this method, we consider the fact that there should be a single possible relation between time-of-death and all other events measured concerning a patient and examine whether MIMIC data conform to this expectation.

Numerical Anomalies. Numerical anomalies stem from two major sources: humans and sensors. Human error is diverse. From typos to inappropriate units-of-measure (UOM). While most errors are sporadic, some may be systematic. For example, putting (and maintaining) sensors on/in patients is a delicate task. Thus, sporadic missing/anomalous values in a continuously measured variable are common. Here again, other patterns are systematic, caused by miscalibrated equipment or improper use [15]. CE contains 2,884 different numerical item IDs. We assume large samples of medical parameters to present a normal (Gaussian) distribution and, thus, follow Ghasemi and Saleh's [16] recommendation to use Z-skewness, Z-kurtosis, and Shapiro-Wilk [17] to test medical data for normality. However, since these tests are all limited to small samples, we use plain kurtosis (kurt) and skewness (skew) and replace Shapiro-Wilk with Anderson-Darling [18] (anderson), that has similar predictive power [19]. To approximate Z-kurtosis and Z-skewness, we add trimmed versions of both (t-kurt, t-skew) where the extreme 10% of values are removed beforehand. We calculate these measures over all numeric item IDs that (1) have >300 CE rows (2) have over five distinct values (3) are not an alarm setting. Of the 790 item-ids remaining, we selected 80 items for which we reviewed known literature regarding extreme values and prepared a database of plausible values. We then count erroneous rows and evaluate the normality tests' predictive power. As a poor man's predictor, we added two simple counting measures, namely #rows and #values, assuming more entries represent more opportunities for error and many different values to indicate abuse of the same field multiple measurement methods. We then performed spearman rank-correlation analysis between these measures and (1) #errors - the raw number of errors, and (2) error rate: the fraction of rows tagged as errors. We also perform cross-correlation between the predictors.

Textual Anomaly Patterns. Textual fields come in two forms. The first is unstructured descriptions, describing the patient's condition, care given, or medical history. In MIMIC-III, most of these descriptions appear in the noteevents table. With no expectations regarding structure, little can be analyzed in terms of correctness. However, data may still be erroneous. For example, case details of one patient may be mistakenly entered into the record of another due to excessive copy-pasting. Weir et al. found that 20% of all notes to contain evidence of copying and 60% to have at least one error [20]. The second type is categorical, recording a selection from a list of options. Optimally, applications contain a closed list of options. However, some categorizations are open-ended, allowing user flexibility, and reducing the load on hospital IT. Flexible fields are noisier. Users may use these fields when there is no prescribed method of entry, or more commonly, when they do not remember the prescribed method. When monitoring a system for data quality, or using its data for research, such fields cannot be monitored effectively and may cause users to use the flexible option rather than the prescribed method, causing gaps/anomalies elsewhere.

2.2   Measuring the Impact of Data Anomalies on Medical Analytics

Visual Analytics. Patient dashboards such as DecisioHealth[2] and Visensia[3], present medical parameters as raw-value graphs or single-number aggregates. For example, such a system can either present BP as a graph showing all recorded BP values, as a single number indicating average BP over the last K measurements or as a colored indicator showing the trend (e.g., red arrow up for a steep upward trend). Unfortunately, all these displays are sensitive to outliers. In our evaluation, we introduce an error found in MIMIC-III to a representative sample of visual analytics and examine the resulting impact on the displays' legibility and usefulness.

---

[2] https://www.decisiohealth.com, Accessed Aug 28, 2019.
[3] http://www.obsmedical.com/visensia-the-safety-index/, Accessed August 28, 2019.

Sepsis Analytics: Robustness of Advanced Warning Systems. Sepsis is a leading cause of death in hospitals worldwide. Early identification of sepsis and the application of appropriate care substantially improve patient outcomes [21]. Due to its importance, numerous measures and prediction models attempt to evaluate whether the patient has sepsis (and its severity) [22], is about to develop sepsis imminently [10], and what is the most effective care path [11]. Researchers have evaluated or developed many of these measures using MIMIC-III, its predecessors, or equivalent datasets. The data used to train the ML model or to assess its prediction power statistically is often cleaned of errors before use. Unfortunately, details of the cleansing procedures and the criteria by which data is classified as erroneous are often omitted. From this large body of work, we chose three representative measures and designed an evaluation to test their sensitivity to anomalous values. Sequential Organ Failure Assessment (SOFA) [23] represents the standard by which other measures are evaluated. SOFA monitors six organ systems: coagulation, hepatic, cardiovascular, respiratory, central nervous system, and renal. Each sub-score ranges between 0 (normal) and 4 (abnormal). To calculate SOFA, we sum sub-scores. The values considered are the most extreme measured on the same day. Organ dysfunction is defined as an increase of at least 2 points. Respiration is measured by PaO2/FiO2. Coagulation is monitored by platelet count. Liver function is determined by blood bilirubin levels. The cardiovascular system function is determined by the existence of hypotension, mean arterial pressure (MAP) below 70 mmHg, or high levels of vasoconstrictors. The central nervous system function is determined by the Glasgow coma score (GCS), a common way to evaluate consciousness. Finally, renal function is evaluated by Creatinine levels. quick-SOFA (qSOFA) [24] represents a light-weight measure that can be applied relatively easily in IT systems and which can be used not only on ICU patients but also on general-ward patients. Designed to estimate the patient odds for developing sepsis and dying from it, this score consists of three parameters: GCS, systolic blood pressure, and respiratory rate. An abnormal value in each adds one to the score, and thus qSOFA ranges between 0–3. A qSOFA score of ≥ 2 indicates organ dysfunction and predicts sepsis. ML-based InSight [10] is trained upon a subset of MIMIC-III. Of the 61,532 ICU stays in MIMIC-III, 22,883 subjects were used. The model uses data from the first 24 h to predict sepsis onset in the next 48. Since InSight's code is unavailable to us, we use a re-implementation trained on the same data. To examine measure sensitivity, we took actual errors from MIMIC-III in parameters used by the measures. Each change was limited to a single error introduced in one parameter, a more likely occurrence than multi-parameter errors or multiple errors in the same parameter. We then calculate the impact of the error by averaging over the score change caused by introducing the error over all ICU-stays. For the ML-model, we only include the InSight cohort.

3 The Impact of Data Quality on Analytics

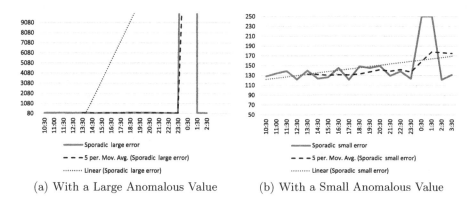

<table>
<tr><td>————— Sporadic large error</td></tr>
<tr><td>— — — 5 per. Mov. Avg. (Sporadic large error)</td></tr>
<tr><td>·········· Linear (Sporadic large error)</td></tr>
</table>

(a) With a Large Anomalous Value

<table>
<tr><td>————— Sporadic small error</td></tr>
<tr><td>— — — 5 per. Mov. Avg. (Sporadic small error)</td></tr>
<tr><td>·········· Linear (Sporadic small error)</td></tr>
</table>

(b) With a Small Anomalous Value

Fig. 1. Effects of anomalies on visual analytics

To motivate the need to teach analytics some common sense about medical parameters, we challenge a set of basic analytics with some values that would be obviously impossible to a casual observer but may appear in a hospital environment. In fact, the values have appeared in the ICU departments of MIMIC-III.

Visual Analytics: Figure 1 contains two examples where random masks a stable BP presented in three popular forms: the raw value (solid blue line), a trend-line (red dotted line), and a moving average over the last five periods (black dashed line). Introducing a common typo present in MIMIC-III (a value of 147147 instead of the intended 147) wreaks havoc on all

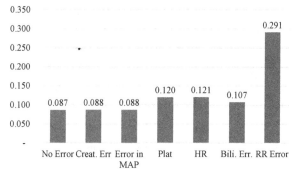

Fig. 2. ML-predictor value average after introducing errors. The baseline values are denoted 'No Error'.

metrics. Auto-adjusted scales would cause the display to become utterly illegible (left). We expect similar effects in single number indicators or colored icons. Even if the display is fixed to the measure's reasonable value range (right), cutting off the erroneous value at 250, the rising trend line and spike in moving average would be manifested as an alarm, wrong-colored indicator or, even worse, masking of the actual trend and changes in the patient status until the data history normalizes.

Sepsis Analytics: Table 2 summarizes our findings. SOFA's directive to take the most extreme value of a period, causes anomalies to have a profound effect ranging between a 2.31 and 3.57 increase in score. As Sepsis is defined to be a change of two points or more in SOFA, any anomaly causes a sepsis onset alarm. Similarly, qSOFA is highly sensitive to single measurements and will cause an immediate alarm. At first glance, the ML-model seems to be more resilient on most measures with

Table 2. Sensitivity of analytics to error

Measure	Parameter	Error used	Score change
SOFA	PaO_2	−17.0	+2.31 (of 15)
	Platelets	0.0	+3.57 (of 15)
	Bilirubin	99,999.0	+3.26 (of 15)
	MAP	−34.0	+2.77 (of 15)
	GCS	None found	NA
qSOFA	GCS	None found	NA
	Systolic BP	−69.0	+1 (of 3)
	RR	2355555.0	+1 (of 3)
InSight	Creatinine	249	0 (of 1.0)
	MAP	−34.0	0 (of 1.0)
	Platelets	0.0	+0.03 (of 1.0)
	HR	86101	+0.03 (of 1.0)
	Bilirubin	99,999.0	+0.02 (of 1.0)
	RR	2355555.0	+0.2 (of 1.0)

score changes ranging over .02-.1, except for RR, which causes a substantial score change even on a single aberrant value. However, upon examining the ML-model's value distribution (Fig. 2) substantial change is evident. In the figure, we present the average for those patients where the affected measure was recorded. Thus, the average baseline probability is 0.087, with a median value of 0.083. However, when we introduce just a single anomalous heart-rate measurement (found in MIMIC-III), the average probability jumps to 0.121.

4   Teaching Analytics Common Sense

Designers that wish to shield their analytics from the anomalies they will eventually encounter when deployed are thus required to teach them some common sense. In this section, we suggest a template for shielding an analytic (Sect. 4.1), a set of anomalous patterns to plan for (Sect. 4.2), and a guide to establishing medical data rules (Sect. 4.3).

4.1   Planning for Common Sense

Figure 3 presents a template designers can follow to endow their analytics with medical common sense. In the first step we identify the data sources. Designers should be aware of the variance in medical systems regarding data collection and storage data and attempt to specify the data required using ontologies and standards such as SNOMED and FHIR, respectively (see book [25]). Step two entails planning for possible anomalous patterns present in each data item by its type. We provide a broad set of patterns in the following section to aid designers in identifying relevant patterns. The third step is to prepare anomaly analysis tools and checklists for implementation and service personnel, allowing them to identify fields and tables that are more prone to error and should be closely monitored. Finally, designers should shield their analytics with multi-faceted

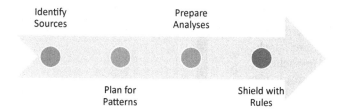

Fig. 3. A Template for teaching analytics medical common sense

rules to monitor incoming numerical data designed to avoid designing for the common cases and remaining susceptible to rare cases (See Sect. 4.3).

4.2 Anomalous Patterns in Medical Data

We now describe anomalous patterns found in MIMIC-III using the analysis methods listed in Sect. 2.1.

Process and Sequence. The existence of a valid identifier for objects and the association of timestamps to events allows process miners (see survey, [26]) to tie together events and reconstruct the care process. Furthermore, ML-based predictors that have a notion of sequence or learn time-based conditions will be sensitive to any variations in the timestamps recorded from the actual ones. We identify four anomalous patterns process/sequence-based analytics must consider.

(1) Missing identifiers. We have recreated the analysis done by Kurniati et al., [13] and found similar levels of missing identifiers in MIMIC-III. Information is missing most prominently from prescriptions (35% of icustay-id's) and noteevents (40% of care giver id's).

(2) Insufficient Resolution. As noted by Kurniati et al. [13], cptevents, noteevents, and prescriptions tables use dates rather than timestamps, severely limiting the ability to use it in conjunction with timestamp level resolution establish cause and effect relationships between events. For example, Sulfonamide, given to patients with bacterial infections, can rarely cause drug-induced reactions such as SJS/TEN [27]. In those cases, symptoms may present as bacterial damage, or of sepsis (caused by the infection). Without timestamps, it is harder to attribute the reaction to the drug or the bacteria. Timestamps associated with medication data may be entered by human loggers and influenced by inaccurate data entry [28]. In addition, a timestamp may refer to when the clinician placed the order, when the drug was administered, or when lab results were reported.

(3) Rounded Times. Human tendency (and at times hospital guidelines) may result in timestamps rounded to the closest hour.

Consider Fig. 4 representing the distribution of CE records per clock-minute. An overwhelming 76% of events are on the hour.

(4) Improbable Time Relations Common sense dictates a single direction between time-of-death and CE events, i.e., CE timestamp and then time-of-death. However, as Fig. 5 shows, this is not always the case. In 415,549 events for 2,679 different patients, CE times were reported after the time-of-death

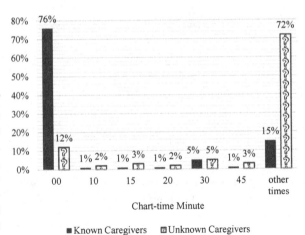

Fig. 4. Distribution of recorded chart-time minute values

recorded in the admissions table. Morbidity analysis and prediction analytics, dependent upon the time before death that a medical action was taken, or a parameter was measured, suffer from such a pattern.

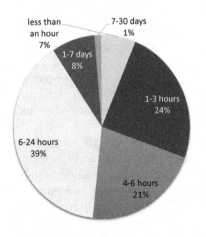

Fig. 5. Report delay in chart events with an **event** time after time-of-death.

Numerical Anomalies. As described above, anomalies may stem from human/ sensor sporadic errors, and systematic discrepancies. Manual examination of MIMIC-III data has identified many such anomalous values, such as leftover default values, typos, inexplicable spikes, and more. Part of an analytic's implementation process is to identify which parameters, in a specific care facility, contain substantial sporadic/systematic error in order to assure an appropriate rule system is deployed or to prioritize such work. The same tools (predictors) can be used to monitor data in order to identify emerging wide-spread anomalies.

Table 3. Identifying parameters with suspicious value distributions.

Predictor	#errors	Error rate
kurt	0.37	0.02
t-kurt	0.04	−0.01
skew	0.16	−0.12
t-skew	0.24	0.17
anderson	0.60	0.10
#rows	**0.80**	0.22
#values	0.64	**0.32**

(a) Spearman correlation between predictors and errors in an item's value distribution.

	kurt	t-kurt	skew	t-skew	anderson	#rows	#values
kurt	1.00	0.31	**0.67**	0.25	**0.89**	0.57	0.44
t-kurt	0.31	1.00	0.15	0.42	0.26	0.03	0.18
skew	0.67	0.15	1.00	0.42	0.61	0.35	0.38
t-skew	0.25	**0.42**	0.42	1.00	0.25	0.18	0.53
anderson	**0.89**	0.26	0.61	0.25	1.00	**0.82**	0.57
#rows	0.57	0.03	0.35	0.18	0.82	1.00	**0.65**
#values	0.44	0.18	0.38	**0.53**	0.57	0.65	1.00

(b) Cross-correlation between the predictors. Bold is most correlated by column.

Table 3a presents the results of correlating our error predictors with actual error over all MIMIC numerical item ids. Surprisingly, the best error predictor by far is #rows with a correlation of 0.8, followed at some distance by #values and anderson. Trimming kurtosis harms its predictive power, while trimmed-skewness is slightly more predictive. Predicting error-rate is a difficult task for this set of predictors. However, in practical applications, predicting the number of errors is probably more useful. With respect to cross-correlation, anderson and kurt are strongly correlated (0.89), as are kurt and skew (0.67). The three measures #rows, #values, and kurt are not perfectly correlated and can be combined using techniques such as regression to create a better predictor.

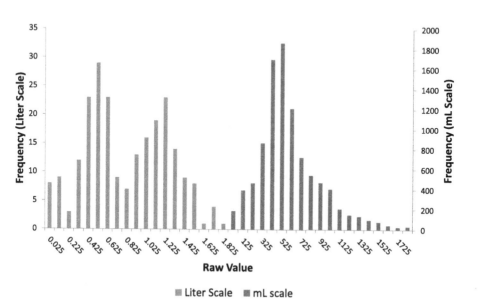

Fig. 6. Tidal volume value distribution.

To exemplify how systematic error can cause a value distribution to deviate from the norm, consider tidal-volume (item id 224685). Applying the predictors above would flag it as having a non-normal distribution since it has a large number of rows (>400K) and distinct values (>4000) and very large kurtosis and anderson values (157,060 and 145,964 respectively). Examining tidal-volume records in MIMIC-III reveals two separate Gaussian distributions, a phenomenon caused by systematic UOM error. In Fig. 6 we present both distributions side-by-side. Normal tidal-volume is 500 ml. Although all tidal-volume entries reported an mL UOM, tidal-volume can be measured by ml or liter (L). We suspect the left-hand distribution (in orange, denoted Liter Scale) has a different intended UOM since it is centered at approximately 0.6. The more common distribution (blue, denoted mL Scale) has a median of 525 ml. This underlying inconsistency would skew aggregate analytics and prediction measures.

Textual Anomaly Patterns. Table 4 presents the top 10 suspicious item ids. Manual evaluation of their value distribution revealed four textual anomaly patterns. Item-id 917 (Diagnosis/Op) is the poster child for *code abuse*, with a hodge-podge of diagnoses and procedures performed on patients and recorded using this field. For example, in row-id 145,199,400 of the CE table, the patient is diagnosed via a 917 item-id with BOWEL OBSTRUCTION. No such diagnosis exists for this patient in its proper location – the diagnoses_icd table, making an automated inference (such as done in [29] impossible. Similarly, a review of the values in

Table 4. Top 10 item ids by distinct value count and the error patterns found

Item id	Label	Value count	Pattern
917	Diagnosis/op	7658	Code abuse (CA)
811	Glucose (70–105)	1394	CA, UOM inline
916	Allergy 1	1090	CA
807	Fingerstick Glucose	1081	CA, UOM inline
446	Micro-Neb Treatment	1015	CA
221	I:E Ratio	791	Unprotected data entry
927	Allergy 2	761	CA
8486	Initials (2 RNs)	412	Hidden identifiers
500	PCA Total Dose	410	UOM inlined

item id 916 found every possible misspelling and variation on a proper name known to data integration specialists. For example, Ace Inhibitors versus A.C.E Inhibitors and Sulfa (Sulfonamides) versus SULFA, sulfa, and Sulfa? Item id's 807 and 811 were commonly used with the more specific type of test (fs/cs) inlined instead of choosing the right item-id. The second typical pattern found is *UOM inline*, where the UOM is added to the value rather than described in the uomvalue field. An alarming case was found in item-id 500 (PCA dose) where the anesthetic dose has a considerable range and several possible units of measurements, making automated identification of anomalous dosages difficult without proper specification of the UOM. Item-id 221 (inspiration-expiration rate) represents an interesting case of *unprotected data entry*. Here, users should enter data in a specific format where the inspiration rate is followed by a colon and then the expiration rate. Without software protections, user error is prevalent in such a field. Item-id 8486 is used in the operational system to denote the initials of the second nurse treating a patient (a similar item id exists for the first nurse). However, initials are not a reliable method of identifying caregivers rendering this data mostly useless in analysis and exemplifying *hidden identifiers*.

4.3 Establishing Tenable Limits on Measured Clinical Parameters

The first challenge to establishing rules for measured clinical parameters is the surprising dearth of accepted guidelines and charts describing the physically possible limits for such measures. Existing articles focus on the normal ranges, such as Flemming et al. [30] for respiratory rate and heart rate in children. Other articles focus on values in disease conditions. For example, Mukaiyama K et al. [31] study the alkaline phosphatase (ALP) values for postmenopausal women. However, there are almost no articles outlining acceptance limits for non-mistake, extreme measured values. Crosfill and Widdicombe [32] present a rare case where the physical limitations of respiratory-related measures are analyzed extensively, not only for humans but for mammalian species in general. Unfortunately, this work is an exception to the rule and stands out as a singular example of such a detailed and well-substantiated examination. Even an exhaustive search was not able to produce a paper or chart describing the humanly possible values of clinical parameters across all human body types, gender, age, and disease variability. In the rest of this section, we review sources of variation and their impact on measured clinical parameters. The first variable to consider is age. An obvious example is the level of estrogen in females, which will be low in childhood, then rise during puberty, and subsequently continue to change dependent upon whether a woman is pregnant or not, and whether menopause has begun, an age-dependent condition as well [33]. Another example is ALP, whose values normally range in children (0–18) between 187.2 and 490.4 IU/L, whereas for adults (19 and above), it ranges between 65.4 and 153.2 IU/L [34]. As an example of limit variances, consider the diversity of maximally possible Heart Rate (HR) at different ages, which can be estimated by $207 - 0.7 \times Age$ [35]. Even this supposedly strict rule cannot be used as-is to serve in the capacity of a limit on plausible values as it is, for example, physically possible for a person aged 28 to present with a maximal HR of 200 if they are of smaller-than-average stature, suffering from a medical condition causing rapid HR, or athletes [36]. Thus, additional variables must be considered. The second variable is gender. There are many parameters whose normative values differ between men and women, for example, the Basal Metabolic Rate (BMR), which is expected to be higher in men than women [37]. A third variable is size. A classic example of parameters that depend on patients' size is drug dosage. In this case, the implications of knowing if a value of an input parameter (from a machine or human record) is true or false are crucial not only for data analysis but for patient safety control systems as well. Administering a large dosage to a small-sized patient can cause intoxication up to death. Conversely, giving an inadequate dosage to a large-bodied patient will inhibit its effectiveness. Finally, the most dominant variable is the comprehensive medical condition of the individual. This condition includes acute and chronic diseases, polymorphisms, and deviations from the norm (such as being a redhead). One specific example is Orthostatic hypotension, which is a condition of very low BP. The mean diastolic BP in this condition is 73.5 mmHg [38], while in MIMIC-III, one can find a few cases with around 30 mmHg, causing difficulty in deciding the cutoff point between extreme albeit rare values and measurement mistakes. Rutan et al. have shown that 12% of adults with Orthostatic hypotension have diastolic BP of 20mmHg [39], making 30mmHg a reliable value for this condition. Conversely, in patients suffering from meningitis caused by gram-negative bacteria, we have found no record of less than 65mmHg and would consider a value of 30mmHg erroneous.

5   Related Work

Weiskopf and Weng [40] suggest a systematic framework comprised of five dimensions of EMR data quality. The analyses presented in this paper fall under the completeness, plausibility, and correctness dimensions. Clifford et al. [41] provide a high-level review of ICU record error sources and the implications on decision support algorithms designed over parameters extracted from this data. In this work, we provide actionable insight into these high-level issues by detailing the anomaly patterns, empirically examining their effect, and providing guidelines to developers on how to shield their analytics from such errors. Maslove et al. [5] analyzed the data quality of five vital signs in MIMIC-III, focusing on prevalent measures that have hourly measurements. In this work we expand the analysis to all clinical parameters, including those collected sporadically. The completeness of the analyses, coupled with the extension to textual values, allows us to and generalize our conclusions to non-ICU settings in which measurement rates are not as frequent. Hauskrecht et al. [42] suggest a detection framework for detect anomalous medical actions (e.g., improper medication prescription or procedure order) based on the patient-state. In this work we are interested in protecting analytics from data errors rather than alerting medical personnel to faulty orders. Ray and Wright [43] propose a system to detect changes in the internal logic of medical decision support system over time. In our work we are interested in the raw data fed into such systems.

6   Conclusions

Visual analytic displays, severity scores, and ML-based analytics should all be tested for their robustness in the presence of data errors, and the appropriate measures should be taken to ensure the safety of patients. We have shown that even a single anomalous value found in MIMIC-III data, when applied on a single parameter, has a severe impact on the effectiveness of analytics. Concerning numerical data, statistical tests are severely limited in most deployment scenarios. Thus, designers must consider the use of rule-based systems to endow their systems with common sense allowing the system to either filter or flag nonsensical data before deriving flawed analytics. Constructing such a rule set must take into account human variation and abnormal conditions rather than normal values. As part of the implementation process, companies are be need identify which fields are more prone to error, in order to prioritize their effort. We have evaluated methods to predict whether a numerical field contains a large number of errors and found #rows, #values, and the Anderson-darling normality test to be the best predictors. Timestamp anomalies and identity errors impact the ability to extract a reliable timeline to ascertain the care plan implemented or compare it to the patient's clinical progression. MIMIC demonstrates several anomalous patterns that can sabotage such an analysis, namely: missing identifiers, insufficient resolution, rounded times, and improbable time relations. Textual fields are used to extract medical terms and to predict/classify patients accordingly. The existence of anomalies or improper use of the fields may impact their ability to construct a robust model of the domain. We have shown a simple method to identify problematic textual fields and the major anomalous patterns to look for when evaluating such fields.

There is much work to be done to create clinical parameter limit guidelines that should consider the inherent variability of the human condition. Such guidelines, if pursued, can serve as a valuable resource for analytics designers and researchers alike. In future work, we hope to validate our analysis by examining the effects of applying common-sense shields using additional public research datasets, both in the ICU and in other clinical settings such as hospital wards, general practitioner's offices, and community clinics. We hope this work will motivate and empower analytics designers to examine their analytics' robustness and endow their systems with common sense using the techniques suggested.

Acknowledgments. This work was performed at GE Healthcare, we wish to thank GE for their cooperation and support. We wish to express our gratitude to Shay Arbiv and Antonio Zaitoun for compiling the analyses presented in this work and to Kaveh Samiee and Hatem Bouabana for providing the Machine-learning-based sepsis prediction model.

References

1. Cooke, C.R., Iwashyna, T.J.: Using existing data to address important clinical questions in critical care. Crit. Care Med. **41**(3), 886 (2013)
2. Badawi, O., Breslow, M.J.: Readmissions and death after ICU discharge: development and validation of two predictive models. PLoS ONE **7**(11), e48758 (2012)
3. Johnson, A.E., Ghassemi, M.M., Nemati, S., Niehaus, K.E., Clifton, D.A., Clifford, G.D.: Machine learning and decision support in critical care. Proc. IEEE. Inst. Electr. Electron. Engi. **104**(2), 444 (2016)
4. Ravì, D., et al.: Deep learning for health informatics. IEEE J. Biomed. Health Inform. **21**(1), 4–21 (2017)
5. Maslove, D.M., Dubin, J.A., Shrivats, A., Lee, J.: Errors, omissions, and outliers in hourly vital signs measurements in intensive care. Crit. Care Med. **44**(11), e1021–e1030 (2016)
6. Li, Q., Mark, R.G., Clifford, G.D.: Robust heart rate estimation from multiple asynchronous noisy sources using signal quality indices and a kalman filter. Physiol. Meas. **29**(1), 15 (2007)

7. Grubbs, F.E., et al.: Sample criteria for testing outlying observations. Ann. Math. Stat. **21**(1), 27–58 (1950)
8. Bowie, M., Begoli, E., Park, B.: Improving quality of observational streaming medical data by using long short-term memory networks (LSTMs). In: 2018 IEEE 34th International Conference on Data Engineering Workshops (ICDEW), pp. 48–53 (2018). https://doi.org/10.1109/ICDEW.2018.00015
9. Johnson, A.E., et al.: MIMIC-III, a freely accessible critical care database. Sci. Data **3**, 160035 (2016)
10. Desautels, T., et al.: Prediction of sepsis in the intensive care unit with minimal electronic health record data: a machine learning approach. JMIR Med. Inform. **4**(3), e5909 (2016)
11. Komorowski, M., Celi, L.A., Badawi, O., Gordon, A.C., Faisal, A.A.: The artificial intelligence clinician learns optimal treatment strategies for sepsis in intensive care. Nat. Med. **24**(11), 1716 (2018)
12. Xiao, C., Choi, E., Sun, J.: Opportunities and challenges in developing deep learning models using electronic health records data: a systematic review. J. Am. Med. Inform. Assoc. **25**(10), 1419–1428 (2018). https://doi.org/10.1093/jamia/ocy068
13. Kurniati, A.P., Rojas, E., Hogg, D., Hall, G., Johnson, O.A.: The assessment of data quality issues for process mining in healthcare using medical information mart for intensive care III, a freely available e-health record database. Health Inf. J. **25**, 1878–1893 (2018). https://doi.org/10.1177/1460458218810760
14. Rumsfeld, J.S., Joynt, K.E., Maddox, T.M.: Big data analytics to improve cardiovascular care: promise and challenges. Nat. Rev. Cardiol. **13**(6), 350 (2016)
15. Kallioinen, N., Hill, A., Horswill, M.S., Ward, H.E., Watson, M.O.: Sources of inaccuracy in the measurement of adult patients' resting blood pressure in clinical settings: a systematic review. J. Hypertens. **35**(3), 421 (2017)
16. Ghasemi, A., Zahediasl, S.: Normality tests for statistical analysis: a guide for non-statisticians. Int. J. Endocrinol. Metab. **10**(2), 486 (2012)
17. Shapiro, S.S., Wilk, M.B.: An analysis of variance test for normality (complete samples). Biometrika **52**(3/4), 591–611 (1965)
18. Anderson, T.W., Darling, D.A.: A test of goodness of fit. J. Am. Stat. Assoc. **49**(268), 765–769 (1954)
19. Razali, N.M., et al.: Power comparisons of Shapiro-Wilk, Kolmogorov-Smirnov, Lilliefors and Anderson-Darling tests. J. Stat. Model. Anal. **2**(1), 21–33 (2011)
20. Weir, C., Hurdle, J., Felgar, M., Hoffman, J., Roth, B., Nebeker, J.: Direct text entry in electronic progress notes. Methods Inf. Med. **42**(01), 61–67 (2003)
21. Rhodes, A., et al.: Surviving sepsis campaign: international guidelines for management of sepsis and septic shock: 2016. Intensive Care Med. **43**(3), 304–377 (2017). https://doi.org/10.1007/s00134-017-4683-6
22. Marik, P.E., Taeb, A.M.: SIRS, qSOFA and new sepsis definition. J. Thoracic Dis. **9**(4), 943 (2017)
23. Vincent, J.-L., et al.: The sofa (sepsis-related organ failure assessment) score to describe organ dysfunction/failure. Intensive Care Med. **22**(7), 707–710 (1996)
24. Singer, M., et al.: The third international consensus definitions for sepsis and septic shock (sepsis-3) consensus definitions for sepsis and septic shock consensus definitions for sepsis and septic shock. JAMA **315**(8), 801–810 (2016). https://doi.org/10.1001/jama.2016.0287
25. Benson, T.: Principles of Health Interoperability HL7 and SNOMED. Springer, Heidelberg (2012). https://doi.org/10.1007/978-1-4471-2801-4
26. Rojas, E., Munoz-Gama, J., Sepúlveda, M., Capurro, D.: Process mining in healthcare: a literature review. J. Biomed. Inform. **61**, 224–236 (2016)

27. Gruchalla, R.S.: 10. drug allergy. J. Allergy Clin. Immunol. **111**(2), 548–559 (2003)
28. Frisch, A., Reynolds, J.C., Condle, J., Gruen, D., Callaway, C.W.: Documentation discrepancies of time-dependent critical events in out of hospital cardiac arrest. Resuscitation **85**(8), 1111–1114 (2014)
29. Wang, S., Li, X., Chang, X., Yao, L., Sheng, Q.Z., Long, G.: Learning multiple diagnosis codes for ICU patients with local disease correlation mining. ACM Trans. Knowl. Discov. Data **11**(3), 1–21 (2017). https://doi.org/10.1145/3003729
30. Fleming, S., et al.: Normal ranges of heart rate and respiratory rate in children from birth to 18 years of age: a systematic review of observational studies. Lancet **377**(9770), 1011–1018 (2011)
31. Mukaiyama, K., Kamimura, M., Uchiyama, S., Ikegami, S., Nakamura, Y., Kato, H.: Elevation of serum alkaline phosphatase (ALP) level in postmenopausal women is caused by high bone turnover. Aging Clinical Exp. Res. **27**(4), 413–418 (2015). https://doi.org/10.1007/s40520-014-0296-x
32. Crosfill, M.L., Widdicombe, J.: Physical characteristics of the chest and lungs and the work of breathing in different mammalian species. J. Physiol. **158**(1), 1–14 (1961)
33. Stevenson, J.C.: A woman's journey through the reproductive, transitional and postmenopausal periods of life: impact on cardiovascular and musculo-skeletal risk and the role of estrogen replacement. Maturitas **70**(2), 197–205 (2011)
34. Eastman, J.R., Bixler, D.: Serum alkaline phosphatase: normal values by sex and age. Clin. Chem. **23**(9), 1769–1770 (1977)
35. Tanaka, H., Monahan, K.D., Seals, D.R.: Age-predicted maximal heart rate revisited. J. Am. Coll. Cardiol. **37**(1), 153–156 (2001)
36. Lester, M., Sheffield, L., Trammell, P., Reeves, T.: The effect of age and athletic training on the maximal heart rate during muscular exercise. Am. Heart J. **76**(3), 370–376 (1968)
37. Kleiber, M.: Body size and metabolic rate. Physiol. Rev. **27**(4), 511–541 (1947)
38. Rose, K.M., et al.: Orthostatic hypotension and the incidence of coronary heart disease: the atherosclerosis risk in communities study. Am. J. Hypertens. **13**(6), 571–578 (2000)
39. Rutan, G.H., Hermanson, B., Bild, D.E., Kittner, S.J., LaBaw, F., Tell, G.S.: Orthostatic hypotension in older adults. the cardiovascular health study. CHS collaborative research group. Hypertension **19**(6_pt_1), 508–519 (1992)
40. Weiskopf, N.G., Weng, C.: Methods and dimensions of electronic health record data quality assessment: enabling reuse for clinical research. J. Am. Med. Inform. Assoc. **20**(1), 144–151 (2013)
41. Clifford, G.D., Long, W., Moody, G., Szolovits, P.: Robust parameter extraction for decision support using multimodal intensive care data. Philos. Trans. R. Soc. A: Math. Phys. Eng. Sci. **367**(1887), 411–429 (2008)
42. Hauskrecht, M., et al.: Outlier-based detection of unusual patient-management actions: an ICU study. J. Biomed. Inform. **64**, 211–221 (2016). https://doi.org/10.1016/j.jbi.2016.10.002
43. Ray, S., Wright, A.: Detecting anomalies in alert firing within clinical decision support systems using anomaly/outlier detection techniques, In: Proceedings of the 7th ACM International Conference on Bioinformatics, Computational Biology, and Health Informatics, BCB 2016, pp. 185–190. Association for Computing Machinery, New York (2016). https://doi.org/10.1145/2975167.2975186

Cdrgen: A Clinical Data Registry Generator
(Formal and/or Technical Paper)

Pedro Alves[1(✉)], Manuel J. Fonseca[3], João D. Pereira[1,2],
and Helena Galhardas[1,2(✉)]

[1] INESC-ID, Lisbon, Portugal
pedro.cunha.alves@tecnico.ulisboa.pt, {joao,hig}@inesc-id.pt
[2] IST, Universidade de Lisboa, Lisbon, Portugal
[3] LASIGE, Faculdade de Ciências, Universidade de Lisboa, Lisbon, Portugal
mjfonseca@ciencias.ulisboa.pt

Abstract. In the health sector, data analysis is typically performed by specialty using clinical data stored in a Clinical Data Registry (CDR), specific to that medical specialty. Therefore, if we want to analyze data from a new specialty, it is necessary to create a new CDR, which is usually done from scratch. Although the data stored in CDRs depends on the medical specialty, typically data has a common structure and the operations over it are similar (e.g., entering and viewing patient data). These characteristics make the creation of new CDRs possible to automate. In this paper, we present a software system for automatic CDR generation, called CDRGen, that relies on a metadata specification language to describe the data to be collected and stored, and the types of supported users as well as their permissions for accessing data. CDRGen parses the input specification language and generates the code needed for a functional CDR. The specification language is defined on top of a metamodel that describes the metadata of a generic CDR. The metamodel was designed taking into account the analysis of eleven existing CDRs. The experimental assessment of the CDRGen indicates that: (i) developers can create new CDRs more efficiently (in less than 2% of the typical time), (ii) CDRGen creates the user interface functionalities to enter and access data and the database to store that data, and finally, (iii) its specification language has a high expressiveness enabling the inclusion of a large variety of data types. Our solution will help developers creating new CDRs for different specialties in a fast and easy way, without the need to create everything from scratch.

Keywords: Clinical data registry · Metamodel · Specification language · Automatic code generation

1 Introduction

In the health sector, the concept of *Clinical Data Registry (CDR)* is crucial [1]. A CDR records data about patients and the health care they receive over time,

P. Alves—This work was developed while the author was a Master student at IST.

and typically is focused on a given medical specialty or a disease. Reuma.pt[1] is an example of a CDR that addresses the Rheumatology medical specialty. Analogously to electronic health record systems, Reuma.pt is an electronic platform that is used by physicians to record personal and medical data (such as demographic, diagnostic and treatment data) related to their patients. CDRs prove to be remarkable for supporting scientific research that addresses clinical problems [3], by providing the data collected in a structured manner.

Physicians often use Excel files to build their own CDR registries, with all the limitations that this process has (e.g., scalability). To overcome limitations, there has been an effort over the years to build more robust CDRs, which are typically composed of three components: (i) a user interface that enables entering and accessing data; (ii) a database to store the entered data; and (iii) a business logic that access the database to fulfill the requests made through the user interface.

In Portugal, where our analysis took place, there have been some initiatives regarding the creation of CDRs, but the existing ones still do not cover all medical specialties. In fact, the twelve CDRs that we were able to find cover only seven medical specialties (see Sect. 2.1). Due to this, new CDRs tend to be created, which implies the development of new software applications from scratch. Since, in general, all CDRs (i) collect data types that are common to various medical specialties (e.g. the patient's name as a string and the birth date as a date); (ii) collects data that covers common topics (e.g., patient's characteristics and medical examinations); and (iii) support user interfaces that offer the same type of functionalities (entering and viewing patient's data), creating a new CDR means a waste of resources because a lot of work is repeated.

To overcome this we could use No-Code Development Platforms (NCDPs) or Low-Code Development Platforms (LCDPs). While on one hand NCDPs[2] allow users to create a new application without writing a single line of code, on the other hand users are not allowed to extend their applications beyond the functionalities provided by the platform. With a little more flexibility than NCDPs, LCDPs [6] are software platforms composed of tools that allow developers to create applications efficiently and with a minimum amount of code that they need to write. In fact, the applications are built and changed in visual environments where the developer can define the underlying data models, business logic, workflow processes and user interfaces. OutSystems[3], Mendix[4] and Appian[5] are some examples of LCDPs.

Although LCDPs offer a more efficient way to create applications at a lower cost than the standard process, they still require some customization effort from the developer. For example, suppose we want to develop several applications that provide the same functionalities but store different data (e.g. CDRs). We would have two options: (i) create the new applications and develop everything

---

[1] http://www.reuma.pt/.
[2] https://www.g2.com/categories/no-code-development-platforms/.
[3] https://www.outsystems.com/.
[4] https://www.mendix.com/.
[5] https://www.appian.com/.

from scratch; or (ii) reuse the first application and change some parts to develop the other applications. In both cases, we would be wasting resources. In the former, we would be repeating much work due to the common aspects of the applications, and in the latter we would have to change the database and also the user interface to accommodate the different data requirements.

In this paper, we describe the design and development of our *Clinical Data Registry Generator (CDRGen)*, which is able to generate a CDR based on an high-level metadata specification (Fig. 1). *CDRGen* interprets the specification and generates the code needed to create and manage the desired CDR. The metadata specification is composed of the data to be collected, the supported types of users and their access permissions. This way, CDRGen enables the creation of CDRs with a reduced effort in terms of software design and development.

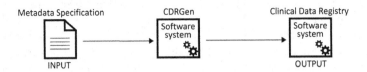

Fig. 1. Overview of the solution to generate a CDR.

Instead of building CDRGen to be able to generate a CDR that covers multiple medical specialties (which, as far as we know, do not happen in an existing CDR), we built CDRGen to be able to generate a CDR for any given medical specialty, as it currently happens in the existing CDRs. We do not follow the former approach, since for each medical specialty could have specific requirements regarding the data collected and the way the data is visualized. Therefore, a CDR that covers multiple medical specialties would be much more complex than one CDR that only covers a given medical specialty. Consequently, building a system, such as CDRGen, to be able to generate a CDR that covers multiples medical specialties would also be more complex.

To develop CDRGen, we performed three steps: (i) identification of the common collected data, functionalities and user interface characteristics of eleven Portuguese CDRs; (ii) syntax definition of the metadata specification language; (iii) creation of an engine to interpret the specification and generate the software logic needed to create and manage the database, the user interface, and the logic that provides the communication between these two components.

To evaluate CDRGen, we used two existing CDRs and measured (i) the *time* required to create a CDR with CDRGen in comparison to the time needed to develop the same CDR without CDRGen; (ii) the *ratio* of tables, their relationships and attributes that exist in the generated database in relation to the expected database based on the data collected by the existing CDR; and (iii) the *ratio* of functionalities that the generated CDR can perform in relation to the functionalities provided by the existing CDR.

The contributions of our paper are threefold: (i) a *metamodel* that describes the common data that must be collected by any CDR; (ii) a *metadata specification language* that enables the developer to specify all the data to be collected, the types of users and their permissions to access the data; (iii) a software application, named *CDRGen*, that parses the metadata specification of the data to be collected by a CDR and generates all the software components needed for the functional CDR.

In Sect. 2 we describe the analysis of existing CDRs, while Sect. 3 describes the resulting metamodel and metadata. In Sect. 4 we present the metadata specification language. Section 5 details CDRGen and Sect. 6 describes its experimental validation. Finally, Sect. 7 presents the conclusions, the limitations and future work.

2   Analysis of Existing CDRs

In this section, we present twelve Portuguese CDRs[6] (Sect. 2.1), identifying the data that they collect and store (Sect. 2.2), the functionalities provided (Sect. 2.3), and their user interface characteristics (Sect. 2.4).

2.1   Portuguese CDRs

We found twelve Portuguese CDRs, which we organize in these three groups:

1. *Specialty-specific*, that cover different medical specialties: (i) Reuma.pt[7] [4]; (ii) Derma.pt[8]; (iii) National Cancer Registry (RON)[9]; and (iv) Information System for HIV/AIDS infection (SI.VIDA)
2. *Cardiological*, from the Portuguese Society of Cardiology (SPC)[10], addressing different diseases of Cardiology: (i) Portuguese Registry of Acute Coronary Syndromes (ProACS); (ii) National Registry of Arrhythmogenic Right Ventricular Myocardiopathy (RNMAVD); (iii) Portuguese Registry on Interventional Cardiology (RNCI); (iv) Portuguese Myocarditis Registry (PMR); (v) Portuguese Registry of Non-compaction Cardiomyopathy (PRNC); and (vi) Portuguese Registry of Hypertrophic Cardiomyopathy (PRo-HCM) [5]
3. *Research* CDRs built in the context of research projects: (i) Umedicine [2]; and (ii) PRECISE Stroke[11].

Table 1 summarizes these CDRs describing, for each of the groups, the concerned CDRs, its medical specialties, the data topics collected, whether the data is structured, and by whom it is accessible. As we can see, all the three groups

---

[6] Although, we were able to find twelve Portuguese CDRs, PRECISE Stroke was not analyzed because we saved it for validating CDRGen.
[7] http://www.reuma.pt/.
[8] http://www.derma.pt/.
[9] https://www.dre.pt/application/file/a/107688306/.
[10] https://www.spc.pt/cncdc/.
[11] https://www.stroke.precisemed.org/home/.

of CDRs collect common data topics, namely regarding patient's characteristics, medical examinations and treatments. In addition, two of the CDR groups collect data with respect to diagnosis, patient's history, patient's follow-up and questionnaires. The data collected is adjusted to the medical specialty of each CDR (e.g., a cardiological CDR collects data specific to diseases of cardiology). All the CDRs are primarily focused on data entry, while data analysis is usually performed by external tools (except for Umedicine, Reuma.pt, RON and SI.VIDA). In six of the CDRs the collected data is structured: it is composed largely by enumerated fields and a few free text fields. We also believe that in the other six CDRs the data is structured, however we were not able to find publicly available information about this. Finally, Reuma.pt and Umedicine are the only CDRs that are accessible both to physicians and patients, while the others can only be accessed by physicians or by other type of health workers.

Table 1. Summary of the Portuguese CDRs. "Unknown" means that we did not find any public information.

Property	Group		
	Specialty-specific CDRs	Cardiological CDRs	Research CDRs
Clinical data registries	- Reuma.pt - Derma.pt - RON - SI.VIDA	- ProACS - PMR - RNMAVD - PRNC - RNCI - PRo-HCM	- Umedicine - PRECISE Stroke
Medical specialties	- Rheumatology - Dermatology - Oncology - Infectiology	- Cardiology	- Urology - Neurology
Collected data topics	- Patient's characteristics - Medical examinations - Treatment - Diagnosis - Disease-specific - Questionnaires	- Patient's characteristics - Medical examinations - Treatment - Diagnosis - Patient's history - Family history - Patient's follow-up	- Patient's characteristics - Medical examinations - Treatment - Symptoms - Patient's history - Questionnaires - Patient's follow-up
Structured data	- Yes (25%) - Unknown (75%)	- Yes (50%) - Unknown (50%)	- Yes (100%)
Accessible to	- Health workers and Patients	- Physicians (75%) - Unknown (25%)	- Physicians, Patients and Clerks

2.2 Collected and Stored Data

Based on the observation of Table 1, we identified six topics that are common across the eleven Portuguese CDRs, to which we call *entity groups*:

- **Person**, representing the patient, namely her characteristics
- **Diagnosis**, representing the symptoms and diseases
- **Treatment**, representing the various treatments that a patient can undergo
- **Questionnaire**, representing the questionnaire that a patient fills in for reporting her health status

- `Medical examination`, representing any medical examination, including laboratory tests
- `Medical appointment`, representing the patient's follow-up in medical visits.

An entity group is composed of *entities*. For example, in a CDR, the `treatment` entity group may contain data concerning a radiotherapy treatment and a chemotherapy treatment. Each of these types of treatments is an entity of the `treatment` entity group. An entity may have one or more instances. For example, an instance of the `radiotherapy` entity describes a particular radiotherapy treatment. Typically, if an entity has several instances, each of them is identified by a date. An entity is composed of *attributes*. For example, in a CDR, the `radiotherapy` entity may be identified by the initial date of a radiotherapy treatment and further characterized by the outcome of that treatment. Each of these two types of data is an attribute of the `radiotherapy` entity. The attributes of an entity have a name (e.g., initial date), a type (e.g., date) and a value (e.g., 01/01/2019). The values of all attributes represent the data that is collected by a CDR. In addition, *constraints* can be defined over attributes (e.g.: a string-type attribute cannot have more than 5 characters; or an attribute cannot hold the null value).

In a CDR, the entity that represents the patient (that we call `patient` entity) always exists and a patient is uniquely identified in the CDR by a value of one particular attribute of that entity. In order to associate the attribute values of any existing entity to a particular patient, in a CDR there are *relationships* between the `patient` entity and the other existing entities. For example, if in a scope of a CDR, a patient can only undergo one radiotherapy treatment, then there is a one-to-one relationship between the `patient` entity and the `radiotherapy` entity. If a patient can undergo various radiotherapy treatments, then there is a one-to-many relationship between these two entities.

2.3   Functionalities

As we stated when we introduced the Portuguese CDRs in Sect. 2.1, the eleven CDRs are primarily focused on data entry, providing *functionalities* that allow the user to add, visualize, edit and delete data. In some CDRs, such as Reuma.pt, and Umedicine, it is possible to search for patients by attributes of the `patient` entity (e.g., patient's identifier or name). In addition, in Reuma.pt and RON, it is possible to generate reports containing details about a patient (e.g., export the inserted data concerning a patient).

Typically, CDRs can be accessed by a *user type* (usually a physician) that is capable of performing all operations over data about any patient. We call `administrator physician` to this user type. In RON and Umedicine, there are other user types, `clerks`, that can only perform a subset of the operations over the data of any patient (e.g., only add personal data of new patients). In Reuma.pt and Umedicine, there is also the `patient` user who can visualize and/or edit a subset of her data (both personal and medical). In addition, in Umedicine there is also the `non-administrator physician` user that can perform all functionalities of the `administrator physician` user but can only add `patient` users.

When generalizing the functionalities and user types, while the administrator physician user can add/visualize/edit/delete data about any patient, the permissions of the other user types may vary. For instance, the patient user may only be able to visualize a subset of her personal data. The UML use case diagram [7] displayed in Fig. 2 presents a generalization of all the user types and functionalities of a CDR whose user types have full permissions. Concretely, there are four different user types: (i) patient; (ii) clerk;

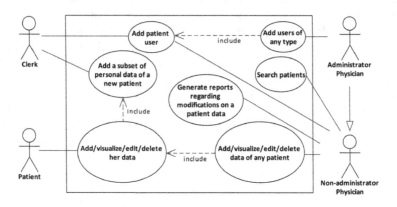

Fig. 2. UML use case diagram that generalizes the user types and functionalities of a CDR.

(iii) administrator physician; and (iv) non-administrator physician. The administrator physician user type always exists for any CDR, while the other three user types may not necessarily exist.

2.4   User Interface

Among the eleven CDRs analyzed, we only had access to the full Umedicine user interface and a few screenshots of Reuma.pt. Nevertheless, this information allowed us to identify the main widgets that are used in their user interfaces to enter data and the type of attribute associated to each widget. For example, a *text box* is used to enter the value of a string-type attribute, while a *dropdown* is used to select predefined values. It is worth noting that the user interface of Umedicine was developed in a previous work of the authors [2], where we used a participatory design approach. This technique involves users in the design decisions of the prototype user interface. In this case, we involved a Urology physician. The resulting Umedicine user interface is composed of data entry forms, to add and edit an instance of a particular entity, and data navigation screens, to visualize data (e.g., all the data of an entity, an overview of all the data from a patient).

Thus, the resulting main data navigation screen is composed at most by six sections, corresponding each to an entity group. Figure 3 shows this screen of a CDR (generated by CDRGen), where we can see the different sections:

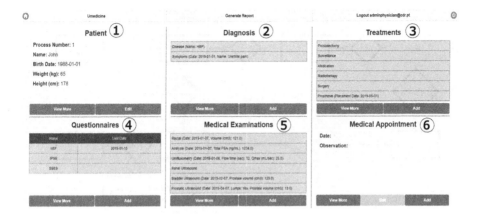

Fig. 3. Screen that allows a physician to get an overview of all data of a patient in a CDR generated based on Umedicine characteristics.

(1) *Patient* represents the **person** entity group; (2) *Diagnosis* corresponds to the **diagnosis** entity group; (3) *Treatments* represents the **treatment** entity group; (4) *Questionnaires* matches the **questionnaire** entity group; (5) *Medical examinations* represents the **medical examination** entity group; (6) *Medical Appointment* represents the **medical appointment** entity group.

3   CDR Metamodel and Metadata

In this section, we start by presenting, in Sect. 3.1, the CDR metamodel based on the analysis performed in Sect. 2. This metamodel describes the metadata of a generic CDR. In Sect. 3.2, we describe an example of a particular CDR.

3.1   CDR Metamodel

Based on the analysis of the eleven CDRs presented in Sect. 2, we propose a metamodel to describe the metadata of a generic CDR. The CDR metadata includes: (i) the types of users supported; (ii) the types of data collected/stored; and (iii) the permissions of the user types over the data. In addition, it also includes the name of the CDR. Figure 4 represents the Entity-Relationship (E-R) model of the CDR metamodel. Rectangles represent entities, ellipses represent attributes while losangles represent relationships. Bold rectangles connected to bold losangles represent weak entities (a dashed underlined attribute name corresponds to the weak entity partial key). A bold edge connecting an entity to a relationship indicates that each entity instance must participate at most once in a relationship instance; if the edge has an arrow pointing to the lonsangle, it means that the entity instance participates in the relationship at most once.

   To describe the metadata of a CDR, as a basis we have the CDR name, the *user types* supported, and the *entity groups* required to define the data collected by a CDR. An entity group has a predefined name (such as person or treatment,

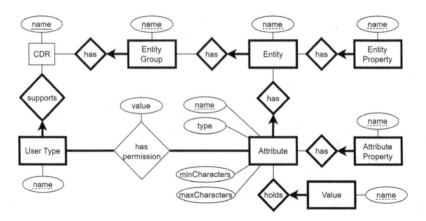

Fig. 4. E-R model of the CDR metamodel.

for example) and is composed of one or more *entities*, as stated in Sect. 2.2. Each entity has a name, can optionally have a set of *properties* and has one or more *attributes*. Each attribute has a name, a type and can optionally have a set of *values*, and properties. The attribute *type* specifies the kind of values the attribute can hold (e.g., string, integer). String and numeric attributes can optionally have a minimum and a maximum number of characters/digits, respectively. A set of *values* is required for enumerated-type attributes. For example, the gender can be an enumerated-type attribute whose values are male and female. A given user type has a certain permission over an attribute. *Permissions* specify if a user type can read and/or write the attribute.

Regarding properties, as an example, an entity can have the *historic* property[12], indicating that the entity can contain several instances that are identified by a date-type attribute. An attribute can have the *not null* property, indicating that it cannot hold the null value.

3.2   A CDR Metadata Example

When the metamodel represented in Fig. 4 is instantiated, it defines the metadata for a particular CDR. In this section, we present an example of such instantiation for the CDR of Example 1.

Example 1. Simple Registry is a CDR that supports two user types: `administrator physician` and `patient`. In *Simple Registry*, we want to collect: (i) the patient's name which, in this case, identifies the patient; (ii) the patient's gender (male or female); and (iii) the date when a patient starts and ends a radiotherapy treatment. In *Simple Registry*, a `patient` user can visualize

---

[12] We call *historic-type entities* to the entities that have the historic property. Otherwise, they are non-historic entities, i.e., entities that can only have one instance (default).

and change her name and gender, but she can only visualize the start and end dates of a radiotherapy.

According to the metamodel represented in Fig. 4, the metadata of the Simple Registry CDR is represented in Fig. 5. The E-R model represented in Fig. 4 is converted into a relational model that consists of relations/tables (e.g., CDR, User Type). Tables have attributes. Arrows represent foreign keys between tables. Each table stores data instances that correspond to the metadata description provided in Example 1. For example, the fact that a patient is characterized by a gender that can take two values is represented by the second tuple in the table Attribute and the two tuples in table Value.

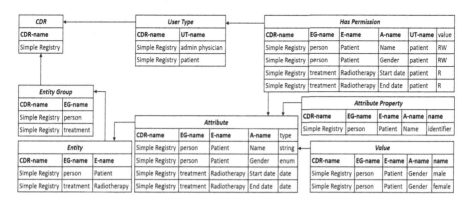

Fig. 5. Metadata of a CDR according to Example 1. Bold fields in the second line of each table represent the primary key of the table.

In Simple Registry there are only two user types that perform a subset of the functionalities available for any CDR, as illustrated in Fig. 6.

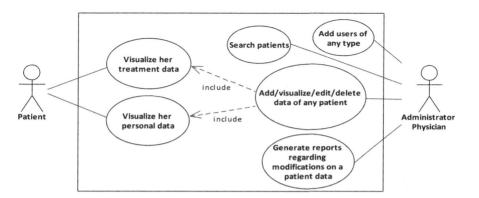

Fig. 6. Use case diagram according to Example 1.

4   Metadata Specification Language

We codified our metadata specification language in the JSON format, because it is sufficiently expressive to represent the metadata in a simple way.

We describe our metadata specification language by following Example 1 (see Fig. 7). The language is composed of the following three keys referring to the JSON root object:

- clinical data registry name (line 2). Its value is a string that represents the name of the CDR (i.e., *Simple Registry*, in this case).
- users (line 3) represents the types of users supported by the CDR. Its value is a string array that can include any of the four user types (except the administrator physician user, since it is always supported).
- entity groups (line 4). Its value is an object that is used to specify the entity groups required to represent the data to be collected by the CDR.

```
 1  {
 2      "clinical data registry name": "Simple Registry",
 3      "users": ["patient"],
 4      "entity groups": {
 5        "person": { "entities": [
 6          { "entity": ["Patient",
 7            { "attribute": ["Name", "string"],
 8              "properties": ["identifier"],
 9              "permissions": { "patient": "RW" } },
10            { "attribute": ["Gender", "enum"],
11              "values": ["male", "female"],
12              "permissions": { "patient": "RW" } }
13          ] } ] },
14        "treatment": { "entities": [
15          { "entity": ["Radiotherapy",
16            { "attribute": ["Start date", "date"],
17              "permissions": { "patient": "R" } },
18            { "attribute": ["End date", "date"],
19              "permissions": { "patient": "R" } }
20          ] } ] } }
21  }
```

Fig. 7. CDR metadata specification according to Example 1.

In the entity groups key value (an object), each key represents an entity group. Since *Simple Registry* only covers two entity groups, there are only two keys: person (line 5) and treatment (line 14). The value of each key is an object that, in this case, only contains the entities key (lines 5 and 14), whose value is an object array that represents the entities of the concerned entity group.

An entity is specified as an object that, in this case, only contains the entity key. The entity key value is an array that must contain at least two elements: one string as the first element and at least one object as the second element. The first element (a string) represents the entity's name. The following objects represent the entity's attributes.

In the specification language, the existence of an entity that represents the patient (i.e., the `patient` entity) is mandatory and must be part of the `person` entity group. Lines 6–13 and 15–20, in Fig. 7, specify the `Patient` and the `Radiotherapy` entities, respectively.

In this example, the object that represents an entity's attribute may contain four keys: `attribute` (mandatory), `properties` (optional), `values` (mandatory and allowed only for enumerated-type attributes) and `permissions` (optional).

The `attribute` key value is an array that must contain two strings. The first string represents the attribute name and the second one represents the attribute type. In this case, among the fourteen attribute types that our system supports, the example specification covers three of them: `string` (line 7), `enum` (line 10) and `date` (lines 16 and 18).

The `properties` key value is a string array that has the attribute properties. For example, the `identifier` property (line 8), indicates that the `Name` attribute of the `Patient` entity is the attribute that identifies the patient in the CDR.

The `values` key value is a string array. The strings represent the enumerated values of an enumerated-type attribute. Line 11 indicates the values of the enumerated-type `Gender` attribute of the `Patient` entity (i.e., male and female).

Finally, the `permissions` key value is an object that specifies the permissions (object values) according to a user type (object key) for the attribute[13]. Lines 9, 12, 17 and 19 exemplify the permissions object, respectively, over the `Name` and `Gender` attributes of the `Patient` entity, and over the `Start date` and `End date` attributes of the `Radiotherapy` entity.

5   CDRGen

This section describes our solution for generating CDRs with a minimum effort in terms of software design and development, called Clinical Data Registry Generator (CDRGen).

The process for generating a CDR starts with a *metadata specification*, that describes the metadata of a particular CDR. *CDRGen* reads this specification, parses it and generates the code needed to produce a *CDR* using a client-server architecture. The resulting CDR is composed of: (i) a *relational database*, that stores the data collected; (ii) a *web user interface client*, through which the user interacts; and (iii) a *web server*, that is responsible for the communication between the database and the web user interface client.

Figure 8 shows the architecture of CDRGen, which is composed of four modules. The *parser* is responsible for reading and parsing a metadata specification, and sending the parsed specification to the code generator container.

The code generator container includes the other three modules: DB generator, Web UI client generator, and Web application generator.

The *DB generator* is responsible for creating a MySQL relational database to store all data that should be collected by the CDR. The database schema is

---

[13] The permissions of the `administrator physician` user type cannot be specified because it has always read and write permissions.

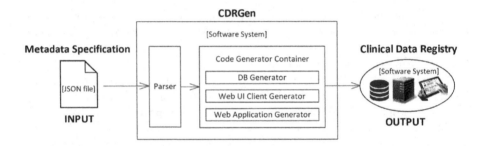

Fig. 8. High-level architecture of CDRGen.

determined by the CDR metadata specification. In short, each entity defined in the metadata specification is represented by a table in the database and each attribute of that entity is represented by an attribute in the same table.

The *Web UI client generator* generates the code that supports the interaction with the user. In particular, this module creates the screens of the user interface.

Finally, the *Web application generator* produces the code regarding the communication between the database and the user interface (i.e., the web server's role). The web server must answer to requests sent by web clients which requires to store and retrieve data from the database.

5.1   Generated CDR User Interface

The user interface of the generated CDR is based on the Umedicine user interface (see Sect. 2.4), and includes data entry forms and data navigation screens.

To give a more detailed example of the screen that displays an overview of all patient's data, suppose that in Simple Registry (Example 1) we had also specified the First Exam and the Second Exam entities in the medical examination entity group. Figure 9 exemplifies what this screen would look like in this case.

Fig. 9. Overview screen of the generated CDR.

As described in Sect. 2.4, this screen is divided in several sections, each representing an entity group defined in the metadata specification. In this example, the specification defines three entity groups, which implies three sections: (1) *Patient* (person entity group); (3) *Radiotherapy* (treatment entity group); and (5) *Medical Examinations* (medical examination entity group). To scale up, a section can occupy 50% of the screen vertically or 100%, depending on the number of specified entity groups. If all six entity groups were specified, then there would be three sections on the top and another three on the bottom of the screen each one occupying 50% of the screen (as shown in Fig. 3).

The content of a section depends on the number of entities belonging to the corresponding entity group. If it has a single entity, then its section displays a set of attributes (name and value) of the single entity. This is the case of the left and middle sections shown in Fig. 9. If the entity is an historic-type entity, the attribute values displayed refer to the most recent entity instance. If an entity group has several entities, its section displays a table, with the name of each entity as a row of the table, as shown in the right section of Fig. 9.

Finally, at the bottom of each section we have buttons. The **View More** button redirects the user to a screen where she can get a detailed vision (instead of an overview) of the entities' data of the concerned entity group. The **Edit** button redirects the user to the data entry form for editing the entity instance in the section. Finally, the **Add** button redirects the user to the data entry form for adding a new instance (i.e., an unfilled form) to an entity.

6   Validation

To validate CDRGen, we compared its process of generating a CDR with the traditional process of building a CDR from scratch in terms of the time spent and the similarity between both CDRs (i.e., the CDR generated using CDRGen and the existing CDR that was built from scratch). We generated two CDRs, PRNC and PRECISE Stroke (see Sect. 2.1) using CDRGen. We chose these two CDRs because we knew what data is collected and the functionalities that they should provide. We used PRNC to test and PRECISE Stroke to validate.

6.1   Metrics

We applied two metrics: *generation time* and *quality*. Generation time represents the time spent by the developer to write the specification of the CDR plus the time that CDRGen takes to generate the CDR. Quality represents the similarity between the generated CDR and the existing CDR. We measured the quality by computing two ratios: (i) the number of tables, their attributes and relationships that the generated CDR's database has in relation to what was expected to have, based on the data that is collected by the existing CDR; (ii) the number of functionalities that the generated CDR supports through its user interface in relation to those provided by the existing CDR.

6.2   Results

For PRNC, the developer spent about 125 min (2 h and 5 min) to write the metadata specification that covered 5 entity groups, 12 entities and 243 attributes.

Then, CDRGen took about 2 s to generate the corresponding CDR. For PRE-CISE Stroke, the developer spent about 284 min (4 h and 44 min) to write the specification that covered 5 entity groups, 23 entities and 633 attributes. Then, CDRGen took about 4 s to generate the corresponding CDR. Concerning the existing PRECISE Stroke CDR, we know that the developing team spent about 300 h to develop it.

For PRNC, CDRGen was able to generate a database with all the tables and relationships between tables that it was supposed to have. Only 1 of the 244 attributes collected by the existing PRNC was not included in the generated database. Concerning the PRECISE Stroke, the generated database had approximately 97% of the tables, 88% of the attributes and 91% of the relationships between tables that it was supposed to have.

Regarding the functionalities of the generated CDRs, we observed that the user is able to perform about 94% and 89% of the functionalities of the existing PRNC and PRECISE Stroke, respectively.

6.3   Discussion

Given that the generation time is in the order of magnitude of seconds, we can conclude that it is irrelevant. Concerning the time to write the metadata specification, for PRNC we took less than half of the time to write the metadata specification than for PRECISE Stroke. This difference was due to the fact that the latter has more than twice the attributes of the former.

With respect to the generated database, we achieved excellent results for PRNC. However, for PRECISE Stroke the results were only good. This happened because in the analysis we made we did not identify some attribute types that we found in this CDR, such as: datetime, time, and image lists where each image has attributes that detail it. We also observed that the historic-type entities that exist in PRECISE Stroke are not properly supported by CDRGen. Some of these attribute types can be easily added to CDRGen, such as time and datetime. In fact, we added these two attribute types to CDRGen after performing the validation with PRECISE Stroke and, once more, we generated a CDR based on the data collected by the existing PRECISE Stroke that already covered these two attribute types. This time, the generated database had approximately 97% of the tables (same as before), 91% of the attributes (instead of 88%) and 91% of the relationships between tables (same as before) that it was supposed to have. The others attribute types require more work but it is possible to extend CDRGen to support them.

With respect to the functionalities provided, we were not able to fully support all functionalities because CDRGen does not support the following three functionalities: (i) visualize statistical data; (ii) export all data of a set of patients; and (iii) delete all data of a patient. However, it is possible to extend CDRGen to generate a CDR with these functionalities.

Overall, as expected, we observe that we were more successful with PRNC than with PRECISE Stroke, since PRNC was used during the analysis. However, even with a CDR not used in the analysis we were able to generate a CDR that supports much of the existing PRECISE Stroke CDR. In fact, although our validation process is not able to fully demonstrate that CdrGen is generic enough to generate any CDR, we are sufficiently confident that CdrGen is capable of supporting the vast majority of requirements of any CDR. We believe in this since CdrGen supports:

- A relatively large set of attribute types (currently, fourteen)
- Several types of relationships between the patient entity and other entities (one-to-one, one-to-many)
- Several types of users whose permissions over the data (read and/or write) can be defined in the metadata specification.

In addition, in both cases, we observe that our solution took less than 5 h to generate a new CDR, which is a great improvement when compared with the time to develop a CDR from scratch. The existing PRECISE Stroke took about 300 h to be developed while with our approach we took less than 2% of this time.

It is still worth noting that, although usability tests have not been carried out in the validation process, the generated user interface was designed based on the Umedicine user interface, which, as previously mentioned in Sect. 2.4, involved a participatory approach with a Urology physician. In this way, the generated user interface was designed in order to have good usability.

7 Conclusions

In this paper, we presented a Clinical Data Registry Generator (CdrGen) capable of generating CDRs with minimal effort in terms of software design and development. Each CDR is generated from a metadata specification that describes, in a high-level language, the characteristics of the CDR to be generated.

We used the existing PRNC to test our CdrGen and then the PRECISE Stroke to validate. The results achieved show that developers can create CDRs in less than 2% of the time it takes with a traditional process. Furthermore, the quality of the generated CDR in terms of the data that it is able to collect and the functionalities provided is above 87% of the expected data and functionalities.

Overall, since the data stored in a existing CDR referring to any medical specialty have common characteristics such as, for example, having a name (e.g., patient's name) and a type (e.g., string), and since the metadata specification that we designed enables to specify these characteristics, then through our metadata specification it is possible to specify the data to be collected by a CDR regardless of the concerned medical specialty. In addition, CdrGen is able to generate, in a more efficient way, a CDR that supports the usual operations that an existing CDR can perform over the data (such as entering and viewing data).

Despite its advantages, CdrGen has some disadvantages that can be clustered into two groups. The first includes the limitations identified during the evaluation of CdrGen using PRECISE Stroke, namely aspects that are not supported (e.g., attribute types and functionalities). The second group includes limitations related to the maintenance of the generated CDR (e.g. change the data to be collected).

For future work we want to support new attribute types and creating the possibility to define attribute types in the metadata specification based on those already supported (e.g., define composed types, such as a list of pairs composed of an integer and a float, for instance). In addition, we want to develop a graphical user interface so that developers can visually specify the metadata of CDRs.

Acknowledgments. This work was supported by national funds through Fundação para a Ciência e a Tecnologia with reference UIDB/50021/2020 (INESC-ID), and UIDB/00408/2020 and UIDP/00408/2020 (LASIGE). The first author would like to thank LAIfeBlood project with reference DSAIPA/AI/0033/2019 for providing him a research grant.

References

1. Santos, M.J., et al.: Reuma.pt contribution to the knowledge of immune-mediated systemic rheumatic diseases. Acta Reumatol. Port. **42**, 232–239 (2017)
2. Lages, N.F., Caetano, B., Fonseca, M.J., Pereira, J.D., Galhardas, H., Farinha, R.: Umedicine: a system for clinical practice support and data analysis. In: Begoli, E., Wang, F., Luo, G. (eds.) DMAH 2017. LNCS, vol. 10494, pp. 102–120. Springer, Cham (2017). https://doi.org/10.1007/978-3-319-67186-4_9
3. Santos, M.J., Canhão, H., Faustino, A., Fonseca, J.E.: Reuma.pt: a case study. Acta médica Port. **29**(2), 83–84 (2016)
4. Faustino, A.: Reuma.pt - the start and the purpose. Acta Reumatol. Port. **43**(1), 6–7 (2018)
5. Cardim, N., et al.: The Portuguese registry of hypertrophic cardiomyopathy: overall results. Rev. Port. de Cardiol. **37**(1), 1–10 (2018)
6. Richardson, C., Rymer, J.R., Mines, C., Cullen, A., Whittaker, D.: New Development Platforms Emerge for Customer-Facing Applications. Forrester Research, June 2014
7. Weilkiens, T.: Systems Engineering with SysML/UML - Modeling, Analysis, Design. MK/OMG Press, February 2008

Prediction of lncRNA-Disease Associations from Tripartite Graphs

Mariella Bonomo, Armando La Placa, and Simona E. Rombo$^{(\boxtimes)}$

Department of Mathematics and Computer Science, University of Palermo,
Palermo, Italy
{mariella.bonomo,armando.laplaca}@community.unipa.it,
simona.rombo@unipa.it

Abstract. The discovery of novel lncRNA-disease associations may provide valuable input to the understanding of disease mechanisms at lncRNA level, as well as to the detection of biomarkers for disease diagnosis, treatment, prognosis and prevention. Unfortunately, due to costs and time complexity, the number of possible disease-related lncRNAs verified by traditional biological experiments is very limited. Computational approaches for the prediction of potential disease-lncRNA associations can effectively decrease time and cost of biological experiments. We propose an approach for the prediction of lncRNA-disease associations based on neighborhood analysis performed on a tripartite graph, built upon lncRNAs, miRNAs and diseases. The main idea here is to discover hidden relationships between lncRNAs and diseases through the exploration of their interactions with intermediate molecules (e.g., miRNAs) in the tripartite graph, based on the consideration that while a few of lncRNA-disease associations are still known, plenty of interactions between lncRNAs and other molecules, as well as associations of the latters with diseases, are available. The effectiveness of our approach is proved by its ability in the identification of associations missed by competitors, on real datasets.

Keywords: lncRNA-disease associations prediction · Tripartite graphs · Decision support

1 Introduction

Long-non-coding RNAs (lncRNAs) are molecules emerging as key regulators of various critical biological processes, and their alterations and dysregulations have been associated with many important complex diseases [9]. The discovery of novel disease-lncRNA associations may provide valuable input to the understanding of disease mechanisms at lncRNA level, as well as to the detection of disease biomarkers for disease diagnosis, treatment, prognosis and prevention. Unfortunately, due to costs and time complexity, the number of possible disease-related lncRNAs that can be verified by traditional biological experiments is very limited. Computational approaches for the prediction of potential

© Springer Nature Switzerland AG 2021
V. Gadepally et al. (Eds.): Poly 2020/DMAH 2020, LNCS 12633, pp. 205–210, 2021.
https://doi.org/10.1007/978-3-030-71055-2_16

disease-lncRNA associations can effectively decrease the time and cost of biological experiments. Computational models quantify the association probability of each lncRNA-disease pair, thus allowing for the identification of the most promising lncRNA-disease pairs to be further verified in laboratory. Such predictive approaches often rely on the analysis of lncRNAs related information stored in public databases, e.g., their interaction with other types of molecules [1,3,10,14]. As an example, large amounts of lncRNA-miRNA interactions have been collected in public databases, and plenty of experimentally confirmed miRNA-disease associations are available as well.

We propose a novel computational approach for the prediction of lncRNA-disease associations (LDAs), based on known lncRNA-miRNA interactions (LMIs) and miRNA-disease associations (MDAs). In particular, we model the problem of LDAs prediction as a neighborhood analysis performed on tripartite graphs in which the three sets of vertices represent lncRNAs, miRNAs and diseases, respectively, and vertices are linked according to LMIs and MDAs. Based on the assumption that similar lncRNAs interact with similar diseases [10], we aim to identify novel LDAs by analyzing the behaviour of *neighbor lncRNAs*, in terms of their intermediate relationships with miRNAs. A score is assigned to each LDA (l, d) by considering both their respective interactions with common miRNAs, and the interactions with miRNAs shared by the considered disease d and other lncRNAs in the neighborhood of l.

Significant predictions for candidate LDAs to be proposed for further laboratory validation are computed by a statistical test performed through a Montecarlo test. The presented approach has been validated on real datasets.

2   Proposed Approach

The main goal of the research presented here is to provide a computational method able to predict novel LDAs candidate for experimental validation in laboratory, given further external information on both molecular interactions and genotype-phenotype associations, but without relying on the knowledge of existing validated LDAs.

The idea of not including any information on existing LDAs in the approach is based on the consideration that only a restricted number of validated LDAs is yet available, therefore a not exhaustive variability of real associations would be possible, affecting this way the correctness of the produced predictions. On the other hand, larger amounts of interactions between lncRNAs and other molecules (e.g., miRNAs, genes, proteins), as well as associations between those molecules and diseases are known, and we have focused our approach on the use of such datasets. In particular, we have considered only miRNAs as intermediate molecules, however the approach is general enough to allow the inclusion of also other molecules in the future.

Problem Statement. Let $\mathcal{L} = \{l_1, l_2, \ldots, l_h\}$ be a set of lncRNAs and $\mathcal{D} = \{d_1, d_2, \ldots, d_k\}$ be a set of diseases. The goal is to return a set $\mathcal{P} = \{(l_x, d_y)\}$ of predicted LDAs.

Let T_{LMD} be a tripartite graph defined on the three sets of disjoint vertexes L, M and D, which can also be represented as $T_{LMD} = \langle (l, m), (m, d) \rangle$, where (l, m) are edges between vertexes in L and M, (m, d) are edges between vertexes in M and D, respectively. In the proposed approach, L is a set of lncRNAs, M is a set of miRNAs and D is a set of diseases. In such a context, edges of the type (l, m) represent molecular interactions between lncRNAs and miRNAs, experimentally validated in laboratory; edges of the type (m, d) correspond to known associations between miRNAs and diseases, according to the existing literature. In both cases, we refer to interactions and associations suitably annotated and stored in public databases.

A commonly recognized assumption is that lncRNAs with similar behaviour in terms of their molecular interactions with other molecules, may also reflect this similarity in their involvement in the occurrence and progress of disorders and diseases [10]. This is even more effective if the correlation with diseases is "mediated" exactly by the molecules they interact with, i.e., miRNAs.

2.1   Scoring of Candidate LDAs

The model of tripartite graph allows to take into account that lncRNAs interacting with common miRNAs, may be involved in common diseases. To this aim, consider two matrixes M_{LL} and M_{LD}. In particular, M_{LL}:

- has h rows and h columns,
- both rows and columns are associated to the lncRNAs in $\mathcal{L}$,
- each element $M_{LL}[i, j]$ with $i \neq j$ contains the number of miRNAs in M linked to both l_i and l_j in T_{LMD};
- each element $M_{LL}[i, i]$ contains the number of edges incident onto l_i.

As for M_{LD}, it:

- has h rows and k columns,
- rows are associated to lncRNAs in $\mathcal{L}$, while columns to diseases in $\mathcal{D}$,
- each element $M_{LD}[i, j]$ contains the number of miRNAs in M linked to both l_i and d_j in T_{LMD}.

We define the *prediction-score* $S(l_i, d_j)$ for the LDA (l_i, d_j) such that $l_i \in \mathcal{L}$ and $d_j \in \mathcal{D}$ as:

$$S(l_i, d_j) = \alpha \left(\frac{M_{LD}[i, j]}{n} \right) + (1 - \alpha) \left(\frac{\sum_x M_{LL}[i, x] \cdot M_{LD}[x, j]}{\sum_x M_{LL}[x, x] \cdot n_j} \right)$$

where $n = \min(M_{LL}[i, i], n_j)$, n_j is the number of miRNAs linked to d_j in T_{LMD}, x are all the possible lncRNA neighbors of l_i and α is a real value in $[0, 1]$ used to balance the two terms of the formula. In particular, the prediction-score measures how much "connected" are l_i and d_j on T_{LMD}, with respect to both the amount of miRNAs they share and the amount of miRNAs that lncRNAs neighbors of l_i share with d_j.

2.2   Prediction of Significant LDAs

Given a set $\mathcal{A}$ of LDAs scored according to the prediction-score computed as described above, it is necessary to select the only associations which are statistically significant, for producing the output predictions. To establish the statistical significance of the considered LDAs, we perform a Hypothesis Test via a Montecarlo simulation [4,5]. The Null Hypothesis is that lncRNAs and diseases have been associated by chance. It is important to focus on the importance that the intermediate miRNAs have in the prediction-score computation and, more in general, in the measure of how much similar is the behaviour of different lncRNAs with respect to the occurrence of diseases. In particular, in the adopted model interactions with miRNAs are the key factors in order to determine the association between a lncRNA and a disease. Let then $(\hat{l}, \hat{m})$ be the pairs in $\mathcal{A}$ and shuffle them for 100 times by producing 100 new sets of pairs $\mathcal{A}_i$. The meaning is to interchange the associations between lncRNAs and miRNAs, still maintaining the same number of interactions. The test to reject the Null Hypotesis consists on comparing the prediction-score $S(l, d)$ of an association (l, d) in $\mathcal{A}$ with the maximum value of prediction-score $\hat{S}(l, d)$ obtained by the same pair in the 100 $\mathcal{A}_i$. The Null Hypotesis is rejected if $S(l, d) > \hat{S}(l, d)$.

3   Results

We have validated the proposed approach on experimental verified data downloaded from starBase [7] for the LMIs and from HMDD [8] for the MDAs, resulting in 114 lncRNAs, 762 miRNAs, 392 diseases and 275 LMIs, 2,201 MDAs. A golden-standard dataset with 183 LDAs has been obtained from the LncRNADisease database [2]. Before proceeding with our discussion, some considerations are needed. Although a number of approaches for LDAs prediction have been presented recently, including machine-learning-based models, only a few of them do not use directly known lncRNA-diseases relationships during the prediction task. However, so far, the experimentally identified known lncRNA-disease associations are still very limited, therefore using them during prediction could bias the final result. Indeed, when such approaches are applied for de novo LDAs prediction, their performance drastically go down [10]. This enforces the idea behind our approach, since neighborhood analysis automatically guides towards the detection of similar behaviours and without the need of positive examples for the training step. With respect to the other approaches which do not use LDAs during prediction (e.g., the p-value based approach in [3]), experimental tests have shown that our approach is able to detect specific situations which are not captured by its competitors. In particular, approaches such as [3] often fails in detecting true LDAs where the lncRNA and the diseases do not have a large number of shared miRNAs. Instead our approach is particularly effective in detecting this kind of situation, since neighborhood analysis allows to detect for example that there are similar lncRNAs associated to that disease.

The proposed approach has been applied to the known experimentally verified lncRNA-disease associations in the lncRNADisease database according to LOOCV. In particular, each known disease-lncRNA association is left out in turn as test sample. How well this test sample was ranked relative to the candidate samples (all the disease-lncRNA pairs without the evidence to confirm their association) is evaluated. When the rank of this test sample exceeds the given threshold, this model is considered in order to provide a successful prediction. When the thresholds are varied, true positive rate (TPR, sensitivity) and false positive rate (FPR, specificity) could be obtained. Here, sensitivity refers to the percentage of the test samples whose ranking is higher than the given threshold. Specificity refers to the percentage of samples that are below the threshold. Receiver-operating characteristics (ROC) curve can be drawn by plotting TPR versus FPR at different thresholds. Area under ROC curve (AUC) is further calculated to evaluate the performance of the tested methods. AUC $= 1$ indicates perfect performance and AUC $= 0.5$ indicates random performance.

We have implemented the p-value based on HyperGeometric distribution for LDAs inference proposed in [3] and compared our approach against it. As a result, the proposed Neighborhoods based approach achieved an AUC equal to 0.67, whereas the p-value based approach scored AUC $= 0.53$, showing that the consideration of indirect relationships between lncRNAs and diseases through neighborhood analysis is more effective.

As for data extracted from StarBase and HMDD, our approach has produced 7, 941 statistically significant LDAs predictions. Among them, it has been able to detect 66 of the 74 verified LDAs of the golden-standard dataset that could have been detected in this larger dataset (due to the presence of lncRNAs and diseases in the golden-standard), 24 out of which not detected by the p-value based approach.

4 Concluding Remarks

We have proposed an approach for LDAs prediction based on neighborhood analysis through a tripartite graph built upon lncRNA-miRNA interactions and miRNA-disease associations. One important fact is that the presented approach predicts potential LDAs without relying on the information of known disease-lncRNA associations. Although many previous study for LDAs prediction use known available LDAs, the latter are still comparatively rare relative to the known lncRNA-miRNA interactions and miRNA-disease associations. Moreover, in the presented research we show that neighborhood analysis performs better than other techniques previously presented in the literature and not based on known LDAs, such as p-value based on HyperGeometric distribution. This is promising and results presented here are to be intended as a first step towards a more complex pipeline, where different types of molecular interactions and associations other than only lncRNA-miRNA will be taken into account (e.g., gene-lncRNA co-expression relationship, lncRNA-protein interactions, etc.). Approaches based on integrative networks have indeed shown to reach better performance, therefore we plan to combine this strategy with the one proposed here on neighborhood analysis. Moreover, taking inspiration from previous studies on social media [6], we plan also to design suitable co-clustering [11,12] and network clustering [13] based methods in order to improve tripartite graph analysis.

Acknowledgements. Part of the research presented here has been funded by the MIUR-PRIN research project "Multicriteria Data Structures and Algorithms: from compressed to learned indexes, and beyond", grant n. 2017WR7SHH, and by the INdAM - GNCS Project 2020 "Algorithms, Methods and Software Tools for Knowledge Discovery in the Context of Precision Medicine".

References

1. Alaimo, S., Giugno, R., Pulvirenti, A.: ncPred: ncRNA-disease association prediction through tripartite network-based inference. Front. Bioeng. Biot. **2**, 71 (2014)
2. Chen, G., et al.: LncRNADisease: a database for long-non-coding RNA-associated diseases. Nucleic Acids Res. **41**, D983–D986 (2013)
3. Chen, X.: Predicting lncRNA-disease associations and constructing lncRNA functional similarity network based on the information of miRNA. Sci. Rep. **5**, 13186 (2015)
4. Giancarlo, R., Rombo, S.E., Utro, F.: Epigenomic k-mer dictionaries: shedding light on how sequence composition influences in vivo nucleosome positioning. Bioinformatics **31**(18), 2939–2946 (2015)
5. Gordon, A.: Null models in cluster validation. In: Gaul, W., Pfeifer, D. (eds.) From Data to Knowledge, Studies in Classification, Data Analysis, and Knowledge Organization, pp. 32–44. Springer, Heidelberg (1996). https://doi.org/10.1007/978-3-642-79999-0_3
6. Ikematsu, K., Murata, T.: A fast method for detecting communities from tripartite networks. In: Jatowt, A., et al. (eds.) SocInfo 2013. LNCS, vol. 8238, pp. 192–205. Springer, Cham (2013). https://doi.org/10.1007/978-3-319-03260-3_17
7. Li, J.-H., et al.: starbase v2. 0: decoding miRNA-ceRNA, miRNA-ncRNA and protein-RNA interaction networks from large-scale CLIP-Seq data. Nucleic Acids Res. **42**, D92–D97 (2013)
8. Li, Y., et al.: Hmdd v2.0: a database for experimentally supported human microrna and disease associations. Nucleic Acids Res. **42**, D1070–D1074 (2014)
9. Liao, Q., et al.: Large-scale prediction of long non-coding RNA functions in a coding-non-coding gene co-expression network. Nucleic Acids Res. **39**, 3864–3878 (2011)
10. Lu, C., et al.: Prediction of lncRNA-disease associations based on inductive matrix completion. Bioinformatics **34**(19), 3357–3364 (2018)
11. Pizzuti, C., Rombo, S.E.: *PINCoC*: a co-clustering based approach to analyze protein-protein interaction networks. In: Yin, H., Tino, P., Corchado, E., Byrne, W., Yao, X. (eds.) IDEAL 2007. LNCS, vol. 4881, pp. 821–830. Springer, Heidelberg (2007). https://doi.org/10.1007/978-3-540-77226-2_82
12. Pizzuti, C., Rombo, S.E.: A coclustering approach for mining large protein-protein interaction networks. IEEE/ACM Trans. Comput. Biol. Bioinform. **9**(3), 717–730 (2012)
13. Pizzuti, C., Rombo, S.E.: Algorithms and tools for protein-protein interaction networks clustering, with a special focus on population-based stochastic methods. Bioinformatics **30**(10), 1343–1352 (2014)
14. Xuan, Z., et al.: A probabilistic matrix factorization method for identifying lncRNA-disease associations. Genes **10**(2), 126 (2019)

DMAH 2020: Invited Paper

Parameter Sensitivity Analysis for the Progressive Sampling-Based Bayesian Optimization Method for Automated Machine Learning Model Selection

Weipeng Zhou[iD] and Gang Luo[(⊠)][iD]

University of Washington, Seattle, WA 98195, USA
{wzhou87,luogang}@uw.edu

Abstract. As a key component of automating the entire process of applying machine learning to solve real-world problems, automated machine learning model selection is in great need. Many automated methods have been proposed for machine learning model selection, but their inefficiency poses a major problem for handling large data sets. To expedite automated machine learning model selection and lower its resource requirements, we developed a progressive sampling-based Bayesian optimization (PSBO) method to efficiently automate the selection of machine learning algorithms and hyper-parameter values. Our PSBO method showed good performance in our previous tests and has 20 parameters. Each parameter has its own default value and impacts our PSBO method's performance. It is unclear for each of these parameters, how much room for improvement there is over its default value, how sensitive our PSBO method's performance is to it, and what its safe range is. In this paper, we perform a sensitivity analysis of these 20 parameters to answer these questions. Our results show that these parameters' default values work well. There is not much room for improvement over them. Also, each of these parameters has a reasonably large safe range, within which our PSBO method's performance is insensitive to parameter value changes.

Keywords: Sensitivity analysis · Automated machine learning model selection · Progressive sampling · Bayesian optimization

1 Introduction

In machine learning, model selection refers to the selection of an effective combination of a machine learning algorithm and hyper-parameter values for a given supervised machine learning task [5]. As a crucial component of automating the entire process of applying machine learning to solve real-world problems, automated model selection is in great need, particularly by citizen data scientists with limited machine learning expertise. In the past few years, many methods, such as Auto-WEKA [8], Auto-sklearn [3], Vizier [1], and AutoGluon-Tabular [2], have been proposed for automating model selection, making it a current hot topic in computer science [5]. Recently, automated methods have

© Springer Nature Switzerland AG 2021
V. Gadepally et al. (Eds.): Poly 2020/DMAH 2020, LNCS 12633, pp. 213–227, 2021.
https://doi.org/10.1007/978-3-030-71055-2_17

outperformed human experts at model selection [1, 2]. Also, multiple major high-tech companies like Google have adopted automated methods as the default model selection approach for building various machine learning models [1]. Despite all of these exciting progresses, a major road blocker still exists in making this wonderful technology widely accessible. Using existing methods to automate model selection on a large data set brings high financial and environmental costs. It often requires using several dozen to several hundred powerful servers continuously for several weeks or several months, incurs a computational cost of several million dollars, and leads to emission of several hundred tons of carbon dioxide due to the large amount of energy consumed [6]. Consequently, researchers and organizations with limited budgets cannot afford using automated model selection to build high-performance machine learning models on large data sets. To address this issue, we developed a progressive sampling-based Bayesian optimization (PSBO) method [9]. It can reduce automated model selection's execution overhead by two orders of magnitude and improve model accuracy at the same time. This is a major progress in green computing and in making automated model selection on large data sets affordable by researchers and organizations with limited budgets.

Our PSBO method has 20 parameters impacting its performance. Each parameter has its own default value that was set empirically in our prior paper [9]. It is unclear for each of these parameters, how much room for improvement there is over its default value, how sensitive our PSBO method's performance is to it, and what its safe range is. In this paper, we perform a sensitivity analysis of these 20 parameters to answer these questions. Our results show that these parameters' default values work well. There is not much room for improvement over them. Also, each of these parameters has a reasonably large safe range, within which our PSBO method's performance is insensitive to parameter value changes.

The rest of the paper is organized as follows. Section 2 reviews our PSBO method. Section 3 describes how we did the parameter sensitivity analysis for our PSBO method. Section 4 shows our experimental results. Section 5 concludes this paper.

2 Review of Our PSBO Method

In this section, we review our PSBO method that has 20 parameters. Table 1 lists these parameters, their definitions, and their default values.

2.1 Overview of Our PSBO Method

Given a data set including multiple features and a prediction target, a set of machine learning algorithms and feature selection techniques, and a hyper-parameter space, we perform model selection to find an effective combination of a machine learning algorithm, a feature selection technique, and hyper-parameter values to build a predictive model with a low error rate. Doing this search often requires testing thousands of combinations. Typical automated model selection methods test each such combination on the whole data set. This causes the search process to take a long time and incur a high computational cost, particularly if the data set is large. To reduce the search time and/or

Table 1. The definitions and default values of the parameters of our PSBO method.

Parameter	Definition	Default value
C_2	The number of cycles of Bayesian optimization done in the second round	3
e	The number of random hyper-parameter value combinations, if any, that are tested for each machine learning algorithm in the first round	20
h_{large}	On a large data set, the number of folds of cross validation done in the final round	3
h_{small}	On a small data set, the number of folds of cross validation done in the final round	10
k	The number of folds of progressive sampling used on a small data set. On a large data set, we take $k = 1$ and use 1-fold progressive sampling	3
L_{f_large}	On a large data set, the maximum number of seconds allowed for doing feature selection when testing a combination of a machine learning algorithm, a feature selection technique, and hyper-parameter values on a fold in the first round of the search process	20
L_{f_small}	On a small data set, the maximum number of seconds allowed for doing feature selection when testing a combination of a machine learning algorithm, a feature selection technique, and hyper-parameter values on a fold in the first round of the search process	10
L_{t_large}	On a large data set, the maximum number of seconds allowed for model training when testing a combination of a machine learning algorithm, a feature selection technique, and hyper-parameter values on a fold in the first round of the search process	20
L_{t_small}	On a small data set, the maximum number of seconds allowed for model training when testing a combination of a machine learning algorithm, a feature selection technique, and hyper-parameter values on a fold in the first round of the search process	10
m	The minimum number of machine learning algorithms, if any, that we try to keep at the end of each round that is not the final round	3
n	The number of new hyper-parameter value combinations chosen for testing in each cycle of Bayesian optimization	10
n_c	For a round that is neither the first nor the last round and a machine learning algorithm, the number of hyper-parameter value combinations that were used in the prior round and are chosen for testing in this round	10

(continued)

Table 1. (*continued*)

Parameter	Definition	Default value
p	The penalty weight given to a combination of a machine learning algorithm, a feature selection technique, and hyper-parameter values that uses feature selection	1.1
q	For a combination of a meta or an ensemble machine learning algorithm and hyper-parameter values, the penalty weight given to every base algorithm used in the meta or ensemble algorithm	0.02
r	The target percentage of machine learning algorithms kept at the end of the first round	40%
s	In the fifth round, for every machine learning algorithm remaining from the prior round, the number of its top hyper-parameter value combinations chosen for testing	10
t	The threshold on the product of the number of data instances and the number of features in the data set for deciding whether the data set is a large or a small one	1,000,000
t_d	For a round that is neither the first nor the last round and a machine learning algorithm, the threshold on the minimum Hamming distance we try to keep among the hyper-parameter value combinations that were used in the prior round and are chosen for testing in this round	2
u	The upper bound on the total number of data instances in the training and validation samples in the fourth round	5,000
w	The target percentage of machine learning algorithms kept at the end of a round that is neither the first nor the last round	70%

the computational cost, researchers have proposed various techniques such as early stopping [4] and distributed computing [7]. These techniques are helpful, but are insufficient for resolving the inefficiency issue. To address this problem, our PSBO method adopts the idea of doing fast tests on small samples of the data set to quickly remove most of the unpromising combinations, and then spending more time on adjusting the promising combinations to come up with the final combination.

More specifically, we do so-called k-fold progressive sampling and proceed in five rounds. In each of the first four rounds, for each fold, we use a training sample to train predictive models and a disjoint validation sample to assess each trained model's error rate. The training sample is initially small and keeps expanding over rounds. At the end of the round, we remove several unpromising machine learning algorithms and shrink the search space. In the fifth round, we identify the best combination of an algorithm, a feature selection technique, and hyper-parameter values to build the final predictive model on the whole data set.

In our PSBO method, we classify the data set as large or small based on whether the number of data instances times the number of features in the data set is over a threshold t. The choice of feature selection technique is treated as a hyper-parameter. We adopt a weight p to penalize the use of feature selection, and a weight q to penalize the use of a meta or an ensemble algorithm. The upper bound on the total number of data instances in the training and validation samples in the fourth round is u. In the following, we describe the five rounds one by one.

2.2 The First Round

In the first round, we start with a small training sample for each fold. For each machine learning algorithm, we test both its default and e random hyper-parameter value combinations, if any, and obtain their error rate estimates. When testing a combination on a fold, we use the algorithm, the combination, and the training sample to build a predictive model and evaluate the model's error rate on the validation sample. During the test, we place a time limit L_f on feature selection and a time limit L_t on modeling training. On a large data set, we set $L_f = L_{f\_large}$ and $L_t = L_{t\_large}$. On a small data set, we set $L_f = L_{f\_small}$ and $L_t = L_{t\_small}$. At the end of the round, we keep the top r (percent) of algorithms having the smallest error rate estimates and remove the other algorithms. If the top r (percent) of algorithms have $<m$ algorithms, we try to keep the top m algorithms, if any.

2.3 The Second to the Fourth Round

In the second round, we increase L_f and L_t by 50%, expand the training sample for each fold, and do four things for each remaining machine learning algorithm. First, we choose n_c hyper-parameter value combinations used in the prior round and test them to obtain their revised error rate estimates. In the selection process, we strike a balance between two goals: a) the selected combinations are away from each other by at least a Hamming distance of t_d, and b) the selected combinations are the ones having the smallest error rate estimates in the prior round. Second, for each combination used in the prior round but not chosen for testing, we multiply its estimated error rate in the prior round by a calculated factor to come up with its revised error rate estimate for the current round. Third, we construct a regression model for estimating a combination's error rate. Fourth, we set $C = C_2$ and do C cycles of Bayesian optimization like that in Thornton et al. [8]. In each cycle, we select n new combinations and test them to obtain their error rate estimates. At the end of the second round, we keep the top w (percent) of algorithms having the smallest error rate estimates and remove the other algorithms. If the top w (percent) of algorithms have $<m$ algorithms, we try to keep the top m algorithms, if any.

The third and fourth rounds work in the same way as the second round, except that C is decreased by one per round.

2.4 The Fifth Round

In the fifth round, we increase L_f and L_t by 50%. For each remaining machine learning algorithm, we choose the top s hyper-parameter value combinations having the smallest error rate estimates in the prior round, if any. Then we do h-fold cross validation on at most u data instances to evaluate each (algorithm, top combination) pair and select the best pair to build the final predictive model on the whole data set. h is set to h_{large} or h_{small} depending on whether the data set is a large one or a small one.

3 Experimental Setup and Procedure

Table 2. The data sets used in the parameter sensitivity analysis for our PSBO method.

Name	No. of classes	No. of training instances	No. of test instances	No. of features
Mammographic mass	2	673	288	6
Car	4	1,209	519	6
Shuttle	7	43,500	14,500	9
Madelon	2	1,820	780	500
Secom	2	1,096	471	591
Arcene	2	100	100	10,000
Waveform	3	3,500	1,500	40
Cardiotocography	3	1,488	638	23
Wine quality	11	3,425	1,469	11
Semeion	10	1,115	478	256
Yeast	10	1,038	446	8
Abalone	28	2,923	1,254	8
KR-vs-KP	2	2,237	959	37
Arrhythmia	16	316	136	279
German credit	2	700	300	20
Diabetic retinopathy debrecen	2	806	345	20
Amazon	49	1,050	450	10,000
Convex	2	8,000	50,000	784
KDD09-appentency	2	35,000	15,000	230
MNIST basic	10	12,000	50,000	784
Dexter	2	420	180	20,000
ROT. MNIST + BI	10	12,000	50,000	784
Parkinson speech	2	728	312	26
Gisette	2	4,900	2,100	5,000
CIFAR-10-small	10	10,000	10,000	3,072
Dorothea	2	805	345	100,000
CIFAR-10	10	50,000	10,000	3,072

In this section, we describe how we did the parameter sensitivity analysis for our PSBO method. Our PSBO method has 20 parameters. Each parameter has its own default value. We used the same 27 data sets adopted in our prior paper [9] (see Table 2) and did our tests on the Hyak computing cluster provided by the University of Washington Information Technology. The cluster runs the CentOS Linux 7.7 operating system and has many computing nodes, each with one 14-core 2.4 GHz Intel Xeon E5-2680 central processing unit and 128 GB memory. We did 20 experiments, one per parameter. In each experiment, we changed the corresponding parameter's value while keeping the other parameters at their default values. For each value tested for this parameter, we ran our PSBO method five times, each with a distinct random seed, for every data set. Each run was given one central processing unit core and 16 GB of memory. At the end of the run, a final predictive model was built on the training data. There are three possible cases for computing the performance measures:

(1) If the parameter is h_{small}, L_{f_small}, L_{t_small}, or k, changing its value impacts our PSBO method's performance on small data sets only. In this case, we computed the average error rate of the final predictive model on the test data and the average search time taken by our PSBO method over all of the runs on every small data set.

(2) If the parameter is h_{large}, L_{f_large}, or L_{t_large}, changing its value impacts our PSBO method's performance on large data sets only. In this case, we computed the average error rate and the average search time over all of the runs on every large data set.

(3) For every other parameter, changing its value impacts our PSBO method's performance on all data sets. In this case, we computed the average error rate and the average search time over all of the runs on each of the 27 data sets.

4 Results

In this section, we present our experimental results. Figures 1, 2, 3, 4, 5, 6, 7, 8, 9, 10, 11, 12, 13, 14, 15, 16, 17, 18, 19 and 20 show the impact of each of our PSBO method's 20 parameters on the average error rate of the final predictive model on the test data and the average search time taken by our PSBO method, one figure per parameter. In each figure, the corresponding parameter's default value is indicated by a vertical bar located near the middle of the figure. The 20 parameters all have reasonably large safe ranges: [2, 3] for C_2, [10, 20] for e, [1, 3] for h_{large}, [6, 12] for h_{small}, [2, 3] for k, [10, 25] for L_{f_large}, [4, 16] for L_{f_small}, [15, 20] for L_{t_large}, [7, 10] for L_{t_small}, [1, 4] for m, [7, 10] for n, [5, 15] for n_c, [1.05, 1.3] for p, [0.015, 0.03] for q, [30%, 40%] for r, [6, 12] for s, [600,000, 1,400,000] for t, [1, 5] for t_d, [3,000, 7,000] for u, and [60%, 70%] for w. For each parameter, its default value is in its safe range, within which our PSBO method's performance is insensitive to parameter value changes. Thus, the 20 parameters' default values all work well. There is not much room for improvement over them.

For each of several parameters such as e, its default value works slightly worse than some other values in its safe range, if we only consider the data sets used in our experiments. Yet, we would not recommend changing these several parameters' default values just because of this, as the opposite case could occur on some other data sets unused in our experiments.

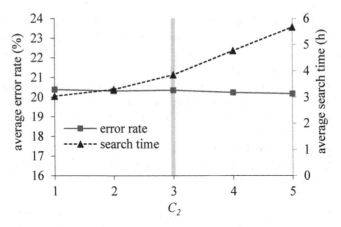

Fig. 1. The average error rate and the average search time vs. C_2.

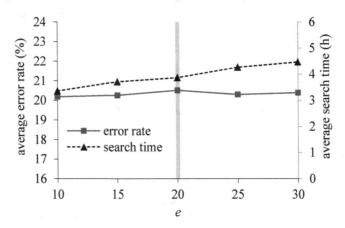

Fig. 2. The average error rate and the average search time vs. e.

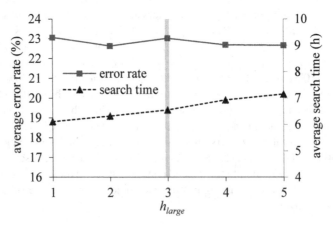

Fig. 3. The average error rate and the average search time vs. h_{large}.

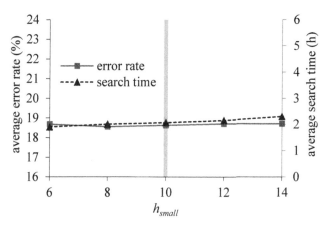

Fig. 4. The average error rate and the average search time vs. h_{small}.

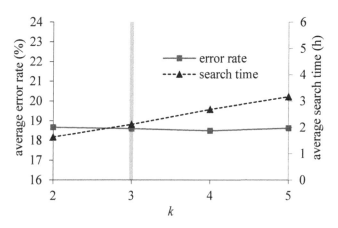

Fig. 5. The average error rate and the average search time vs. k.

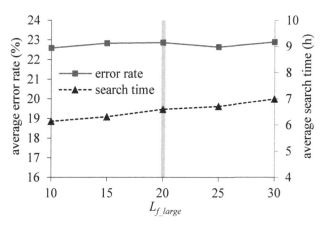

Fig. 6. The average error rate and the average search time vs. L_{f_large}.

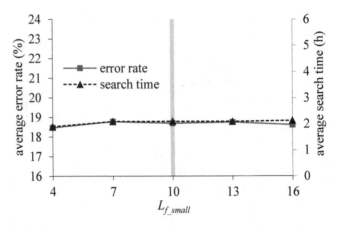

Fig. 7. The average error rate and the average search time vs. L_{f_small}.

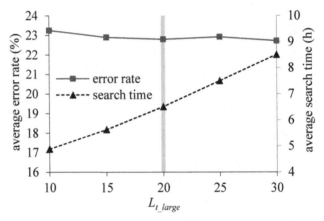

Fig. 8. The average error rate and the average search time vs. L_{t_large}.

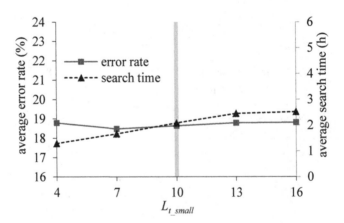

Fig. 9. The average error rate and the average search time vs. L_{t_small}.

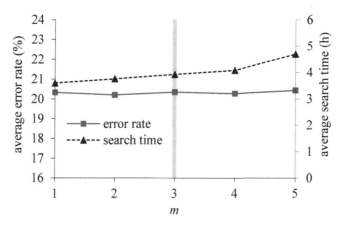

Fig. 10. The average error rate and the average search time vs. m.

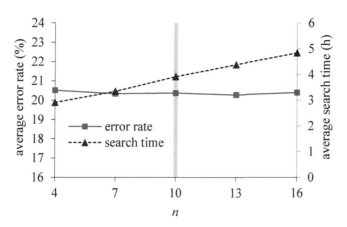

Fig. 11. The average error rate and the average search time vs. n.

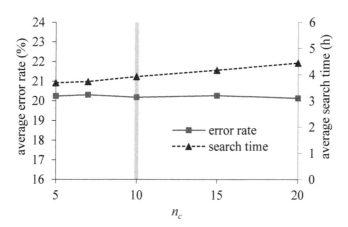

Fig. 12. The average error rate and the average search time vs. n_c.

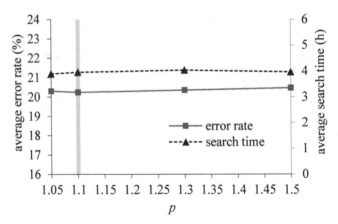

Fig. 13. The average error rate and the average search time vs. *p*.

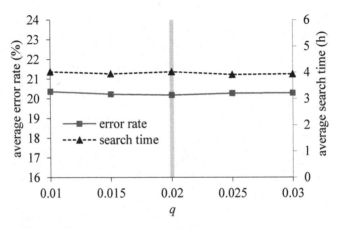

Fig. 14. The average error rate and the average search time vs. *q*.

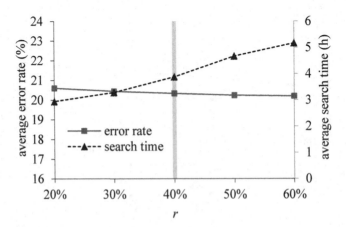

Fig. 15. The average error rate and the average search time vs. *r*.

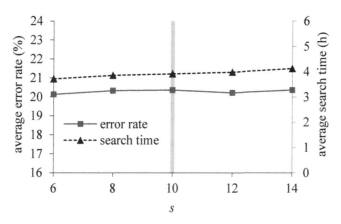

Fig. 16. The average error rate and the average search time vs. *s*.

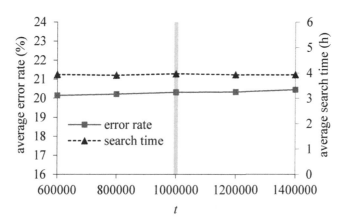

Fig. 17. The average error rate and the average search time vs. *t*.

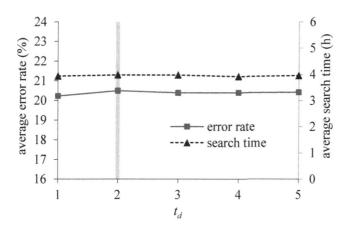

Fig. 18. The average error rate and the average search time vs. t_d.

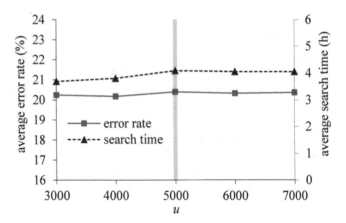

Fig. 19. The average error rate and the average search time vs. *u*.

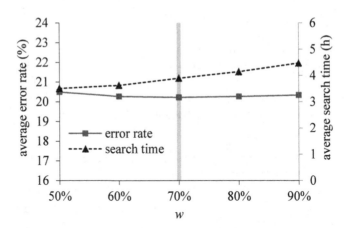

Fig. 20. The average error rate and the average search time vs. *w*.

5 Conclusion

Our sensitivity analysis shows that the default values of our PSBO method's parameters work well. There is not much room for improvement over them. Also, each of these parameters has a reasonably large safe range, within which our PSBO method's performance is insensitive to parameter value changes. These results fill a gap left by our prior study [9].

Acknowledgment. GL was partially supported by the National Heart, Lung, and Blood Institute of the National Institutes of Health under Award Number R01HL142503. The funders had no role in study design, data collection, and analysis, decision to publish, or preparation of the manuscript. This work was facilitated through the use of advanced computational, storage, and networking infrastructure provided by the Hyak supercomputer system at the University of Washington.

Authors' Contributions. WZ did the computer coding work, conducted the experiments, participated in doing literature review and designing the study, and wrote the paper's first draft. GL conceptualized and designed the study, participated in doing literature review, and rewrote the whole paper. Both authors read and approved the final manuscript.

References

1. Golovin, D., Solnik, B., Moitra, S., Kochanski, G., Karro, J., Sculley, D.: Google Vizier: a service for black-box optimization. In: Proceedings of 23rd ACM SIGKDD International Conference on Knowledge Discovery and Data Mining, pp. 1487–1495. ACM Press, New York (2017)
2. Erickson, N., et al.: AutoGluon-Tabular: robust and accurate AutoML for structured data. arXiv preprint arXiv:2003.06505 (2020)
3. Feurer, M., Klein, A., Eggensperger, K., Springenberg, J., Blum, M., Hutter, F.: Efficient and robust automated machine learning. In: Proceedings of Annual Conference on Neural Information Processing Systems 2015, pp. 2944–2952 (2015)
4. Li, L., Jamieson, K.G., DeSalvo, G., Rostamizadeh, A., Talwalkar, A.: Hyperband: a novel bandit-based approach to hyperparameter optimization. J. Mach. Learn. Res. **18**(185), 1–52 (2017)
5. Luo, G.: A review of automatic selection methods for machine learning algorithms and hyperparameter values. Netw. Model. Anal. Health Inf. Bioinf. **5**(1), 1–16 (2016). https://doi.org/10.1007/s13721-016-0125-6
6. Strubell, E., Ganesh, A., McCallum, A.: Energy and policy considerations for deep learning in NLP. In: Proceedings of 57th Conference of the Association for Computational Linguistics, pp. 3645–3650. Association for Computational Linguistics (2019)
7. Swearingen, T., Drevo, W., Cyphers, B., Cuesta-Infante, A., Ross, A., Veeramachaneni, K.: ATM: a distributed, collaborative, scalable system for automated machine learning. In: Proceedings of 2017 IEEE International Conference on Big Data, pp. 151–162. IEEE Press, New York (2017)
8. Thornton, C., Hutter, F., Hoos, H.H., Leyton-Brown, K.: Auto-WEKA: combined selection and hyperparameter optimization of classification algorithms. In: Proceedings of 19th ACM SIGKDD International Conference on Knowledge Discovery and Data Mining, pp. 847–855. ACM Press, New York (2013)
9. Zeng, X., Luo, G.: Progressive sampling-based Bayesian optimization for efficient and automatic machine learning model selection. Health Inf. Sci. Syst. **5**(1), 2 (2017).

Short Paper

Extended Abstract: Programming Heterogeneous Data Applications with Knowledge Graphs

Michael Cafarella

MIT
michjc@csail.mit.edu

Abstract. Heterogeneous data applications pose an urgent social need. If data systems could better manage heterogeneous data, they could answer crucial questions such as, "Which authors have written the most-cited papers that mention both Remdesivir and a filovirus?" Or, "How much will COVID-related spending raise the federal debt?" Knowledge Graphs (KGs) such as Wikidata have shown remarkable success in incrementally building high-quality heterogeneous datasets, and could potentially answer some of these questions. But the number of KG-driven applications is very small. We outline our efforts to build a *knowledge graph programming system*, whch makes building heterogeneous KG-driven data applications cheaper and easier.

Knowledge Graphs have an established track record for yielding high-quality heterogenous data stores. For example, English-language Wikidata contains more than 89M entities and more than 1.1B facts, yet still retains remarkable quality when it comes to deduplicated objects, consistent property names, and so on. Knowledge Graphs hold the promise of flexibly storing heterogeneous data at a very large scale. Unfortunately, KG-driven applications include structured Web search, voice assistants, and little else.

One explanation is that KG application development is time-consuming, tedious, difficult, and expensive, with only the best-resourced technical companies able to manage the challenge. In response, we propose a novel *knowledge graph programming system* that makes KG-driven app development easier, faster, and cheaper. Our planned system has three core and interlocking elements.

1. **Built-in Data Libraries** — Our system sidesteps the need for manual data loading by building common datasets into the language. A modern programming language always comes with certain data structure libraries, such as arrays and sorted trees. The same should be true of standard datasets. Doing so will avoid the need for manual ingestion code. It will also help different programmers converge on naming and defining standard datasets, a common problem seen by anyone who has tried to reproduce an experiment on "last night's query log" or "the most recent trained model." As a result, using external datasets should become faster and collaboration among developers should be much easier.

   The closest approximation we have for a "standard" dataset are popular knowledge graphs such as Wikidata, MusicBrainz, and others. As a result, our

V. Gadepally et al. (Eds.): Poly 2020/DMAH 2020, LNCS 12633, pp. 231–232, 2021.
https://doi.org/10.1007/978-3-030-71055-2

system will automatically map KG-related values, types, and datasets into corresponding language structures. For example, the Wikidata entity for `Barack Obama` can be used as a simple value in just one line of code. The `Human` entity can be imported as a type. The United States' `population` data can be imported as a table. For organization-specific datasets like "last night's query log," the programming system can use the organization's in-house knowledge graph.

2. **Data Sharing as a First Class Operation** — Data applications are sometimes thought of as a "pipeline" of steps, which eventually yields a refined dataset. But almost every data pipeline ingests data from a variety of external sources, and most of them export data to users, or colleagues, or external databases, and so on. The data pipeline in reality is more accurately thought of as a tiny portion of a much larger — though implicit — data collaboration graph. The invisible edges in this graph are made up of email attachments, transient database loads, and swapped files.

Unfortunately, by keeping sharing relationships implicit, app developers cannot easily debug data quality across sharing boundaries. If an app emits an incorrect value due to a flaw in an upstream step in this implicit collaboration graph, there is effectively no way for the developer to fix the root cause. At best, the developer can apply a local patch that must be reapplied every time the remote input changes. The patch must often be rewritten. This is another way in which standard KG-driven application development is expensive and slow.

Our system will get all of its inputs and outputs via a standard publication mechanism. Sharing data with a colleague will be as simple as sending a URL, then using it in application code. This will make the data collaboration graph more explicit, making upstream errors easier to collaboratively repair.

3. **Automatic Provenance** — Building a data application largely consists of debugging data quality problems. As a result, high-quality provenance information is crucial. Provenance systems are common in practical analytical deployments, but they tend to be custom-built rather than based on standard software. As a result, building a provenance system is time-consuming and expensive, and thereby limited to the largest and best-resourced organizations.

We will execute user code in a special runtime container that automatically generates provenance data for every execution. Every value and variable will receive a recorded unique identifier, so it can be examined by downstream users. One challenge with automatic provenance collection is that the resulting provenance graph can overgenerate dependency links. Dedicated provenance systems with semantic knowledge of the operations can reduce this overgeneration, but also require the developer to modify her code. We will instead build automatic provenance capture that allows for optional "knowledge annotations" of the provenance graph. These will allow administrators with semantic insight into particular operations to trim back the overgenerated graph. As this programmatic knowledge base grows, the resulting automatic provenance graphs will become more useful.

Author Index

Alves, Pedro 188
Athanasiadou, Rodoniki 141

Barmpis, Konstantinos 54
Begoli, Edmon 72
Bergman, Ruth 171
Bonomo, Mariella 205

Cafarella, Michael 3, 66, 231
Chen, Gang 100
Christian, J. Blair 87
Coyle, Linda 87

DeWitt, David 3
Doherty, Jennifer 87
Dong, Qifei 151
Durbin, Eric B. 87

Fonseca, Manuel J. 188
Friedman, Bruce 171

Gadepally, Vijay 3, 66
Galhardas, Helena 188
Geissmann, Isabel 25
Gulum, Mehmet A. 120

Hansen, Nils 25
Hara, Takahiro 37
Hwang, Hyunseung 81

Kantardzic, Mehmed 120
Kepner, Jeremy 3
Klasky, Hilda B. 87
Kolovos, Dimitris 54
Kozyrakis, Christos 3
Kraska, Tim 3

La Placa, Armando 205
Lengweiler, David 25
Li, Xin 100
Luo, Gang 213

Mahbub, Maria 72
Medhat, Fady 54

Neubauer, Patrick 54

Paige, Richard F. 54
Penberthy, Lynne 87
Pereira, João D. 188
Philipp, Sebastian 25
Price, Benjamin 66

Razavian, Narges 141
Rezig, El Kindi 66
Rombo, Simona E. 205

Sagi, Tomer 171
Schönholz, Jan 25
Schuldt, Heiko 25, 42
Shmueli, Nitzan 171
Srinivasan, Sudarshan 72
Stanley, Christopher 87
Stiemer, Alexander 25, 42
Stonebraker, Michael 3, 66
Störl, Uta 42
Stroup, Antoinette 87

Tong, Yao 100
Tourassi, Georgia D. 87
Trombley, Christopher M. 120

Vanterpool, Allan 66
Vogt, Marco 25, 42

Wang, Yue 151
Whang, Steven Euijong 81
Wu, Xiao-Cheng 87

Yan, Hang 100
Yoon, Hong-Jun 87

Zaharia, Matei 3
Zhang, Xiaoyi 141
Zhang, Zhenxiang 100
Zhou, Weipeng 213
Zolotas, Athanasios 54

Printed in the United States
By Bookmasters